BADGERLANDS

Also by Patrick Barkham

The Butterfly Isles

BADGERLANDS

The Twilight World of Britain's Most Enigmatic Animal

Patrick Barkham

GRANTA

Granta Publications, 12 Addison Avenue, London W11 4QR

First published in Great Britain by Granta Books 2013

A CIP catalogue record for this book
is available from the British Library

9 8 7 6 5 4 3 2 1

ISBN 978 1 84708 504 7

www.grantabooks.com

Typeset in Garamond by M Rules

Printed and bound by CPI Group (UK) Ltd, Croydon, CR0 4YY

For my mum, Suzanne

CONTENTS

1

Black-and-White

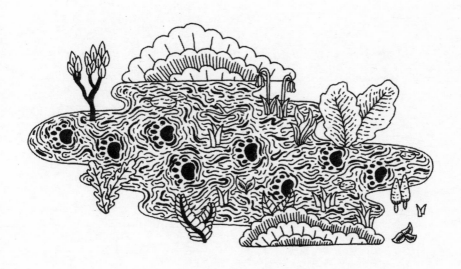

There was a thin film of ice on the car windscreen when I stepped outside the farmhouse at a quarter to ten. The sky was mad with stars and the bare branches were bullied by the wind that blew in from the east. The hillside was dusty black, its hedges a darker line of pencil; at the top blinked the four red lights of a tall transmitter on the heights of the Mendips; lower down, a lane snaked around a wooded bulge in the slope, a belt around an expanding belly. A solitary pair of headlights swept between the wintry hedges and then disappeared over the horizon.

I had ventured out on a bleak March night to look for badgers. With its reputation for sorcery and subterranean dwellings, Wookey Hole seemed a particularly good place to begin. I was seeking badgers not in the village's famous caves, home to a vaguely human-shaped stalagmite called the Witch of Wookey Hole and now an amusement park, but on a small farm just outside the village in Somerset. The farm was owned by Nick Lee, a short, smiley man with strong fingers and deep wrinkles that gathered shrewdly around his eyes. He had sold his dairy cows a few years ago and badgers, widely blamed by many who work the land for spreading bovine TB among cattle, had something to do with it. Now, Nick and his wife Sue got by on a bit of everything: beef cattle, bed-and-breakfast, camping. They also tolerated a couple of badger setts on their land and told me where to find them.

One of the less well-known caves above the Wookey ravine is called Badger Hole. The badger, and its other old names – brock, pate, grey, bawson, billy, black-and-white – are written into our landscape: Badger in Shropshire, Brocklebank in the Lake District, Grayswood in Surrey, Badgers Mount in Kent, Broxbourne in Hertfordshire. At least 140 Anglo-Saxon place names originate from *broc*. There are a fair number of fox-related place names but, apart from the odd hamlet called Rattery and a brace of Pigeon Lanes, few other species have such a strong linguistic presence in human habitations. We might imagine this shows the esteem in which we hold this most independent of wild animals, but the history of our dealings with the badger over the centuries is one of relentless brutality. This, too, is revealed in our language, in the origin of the verb 'to badger'.

Our country has not merely been named after badgers; it has been shaped by them. Badgers' spectacular earthworks have changed the lines of hedgerows, turned pasture into woodland, and forced farmers to abandon or at least reconsider the way they farm their land. A higher density of badgers lives in Britain than anywhere else in the world. The badger is our biggest surviving carnivore, although, like us, it is omnivorous, and there are plenty of vegetarian badgers. Over the centuries, we have driven more formidable species – bears, wolves, lynx – to extinction, but the badger has endured. Perhaps it is because we are so oblivious to these sylvan animals living quietly all around us.

This is odd because the badger is an unmistakable beast, the very opposite of a camouflaged creature. The size of a spaniel that has been completely reshaped, it has supremely powerful claws, the hulking neck of a body-builder and, of course, that fright mask: the long white face burnished by two black stripes. Badger hair is actually white and only the tip of its guard hairs, which form a protective outer layer, is black. This gives it a grey appearance but its face badge shows up like a beacon in the gloom of a wood at night, long after it has become impossible to pick out a fox or deer. And yet it is astonishing how rarely we see a badger, and how little most of us know about its habits and natural history.

At various points in my life I had passed through badgery places such as Brockley and Pately Bridge; I read about the adventures of Mr Badger in *The Wind in the Willows* as a child, and as an adult drank a pint in the Fat Badger pub in West London, now sadly replaced by a pizzeria; I had followed those brown road signs adorned with the badger's mask that signify a nature reserve; bid for a yellowed stuffed badger on eBay, and mooched past their underground setts; but, like so many people, I had barely seen a wild badger. Dead ones, of course, shunted and rolled into the gutter, stumpy legs at wonky angles after they had stubbornly faced down an onrushing car and come off second best. Plenty of lifeless badgers. But in all my years growing up in the countryside, I had caught a glimpse of a live badger only at odd moments when I had taken a turning away from normal life and entered a strange, nocturnal universe. Badgerland.

Until I began to seek out the parallel world populated by badgers, my lifelong record had been three fleeting sightings. Once, bleary-eyed, I hailed a taxi to escape the mayhem of Glastonbury Festival and saw a low-slung shape cross the lane as we drove through the Polden Hills. The first live badger of my life, at the age of twenty-nine. Another time, an accident blocked the M1 and I veered off at a random junction and meandered, lost, along a B-road in Northamptonshire. A badger was picked out in my lights as it made its nightly perambulation. A third I stumbled upon in daylight when my dad and I were traipsing knee-deep through a hazardous bog on Dartmoor. The dead bracken shook in front of us as a badger – resting up outdoors in the late afternoon – dashed away with unexpected speed.

My grandma would have been disappointed that I had reached the middle of my life and spent so little time with badgers. Jane Ratcliffe, my mum's mother, was obsessed with them. She had a skull on her sideboard and a special badger gate in the dry-stone wall between her garden and the wood. With her large glasses and white hair, Grandma looked a bit like an owl, and even more so when she spent long evenings perched up a tree, watching badgers pootle about on the woodland floor below. Much of her adult energy was poured into rehabilitating wild animals. She nursed badgers back to health, gradually moving them from garage to artificial garden sett and then back into the wood. They ignited such passion that she wrote a book about them. It was called *Through the Badger Gate* and it read like a love letter to Bodger, her first badger. By the time I was born, Grandma's badgers were back in the wild, or dead, and the cages in her garden were full of injured tawny and barn owls instead. As a child, I was not aware that my grandma had played a crucial role in changing the relationship between humans and badgers in Britain, perhaps for good.

Unlike my childhood, bereft of all but fictional badgers, every evening was badger night for the residents of Wookey Hole. Nick and Sue's daughter, Jess, remembered what passed for entertainment in these parts during her teenage years. One day a young man stuffed a firework inside a dead badger's head and they all stood back and watched the explosion. 'Everyone hates running over badgers ...' Jess explained. I nodded empathically, pondering the heartbreak of accidentally flattening a badger. '... Because it fucks up your car.'

This was the first jolt upon entering Badgerland. Watching badgers in the wild, I hoped, would be a step towards understanding these unique animals, and our complex, contrary relationship with them. In recent years, however, the land where badgers roamed had become a battleground on which collided all kinds of arguments about the disjuncture between the town and the country, the crisis in farming, the rights of animals and how we should best live alongside wild creatures in our countryside. What I had not yet realised was the compelling wild power of the animals I would eventually discover in the many different Badgerlands. I was also completely unprepared for the passion of the people who set up home in this dusky underworld, bearing both ancient enmities and a more contemporary sentimentality towards badgers.

I hoped to trace the strange history of our relationship with badgers and find out why it was so vexed; why some people devoted their lives to feeding badgers while others risked prison to torture them; and why what was styled as a cultural war between the countryside and the city was being waged through the badger, a ferocious debate over the animal's contribution to bovine TB in cattle and whether that merited a cull of badgers.

We divide most wild animals in Britain into pests or national treasures. A few, such as the fox, are both. There is clarity in the human relationship with the fox; we either despise or admire it, usually for the same reason: that it is a cunning, adaptable predator. Our relationship with the badger is far less cogent, and much more mysterious. For all the love lavished on badgers since Kenneth Grahame made Mr Badger the paternal rock upon whom less reliable species could depend, there is an older tradition of massacring these mammals. The badger may be popularly seen as a noble, quintessentially British beast in its tenacity, determination and defence of its fortress-like home, and yet it has endured centuries of persecution probably more brutal than has been meted out to the fox. Unlike the fox, we don't quite know what to make of the badger. Is she stupid or intelligent? Predator or prey? Tame pet or wild thing? Fearsome or fearful? Good or evil?

*

I tiptoed silently across the field, away from our jolly conversation inside the warm farm kitchen, obeying the first law of badger watching: keep downwind of the sett. I was woefully equipped for the night, and not just because I'd borrowed a torch that threw out the yellowy light of a guttering candle. It was my first night out alone in the dark, the proper pitch black that could only be delivered by the countryside, for years. I had little knowledge of the night, and less experience of it. Like most adults, I banished darkness; my life was constantly lit up, as if I was afraid of something. Without being regularly tested, our senses become even dimmer. Now, lacking the comfort of clear sight, my usually underemployed ears and nostrils strained. In the chill at the end of a long winter, there was no smell. What could I hear? Only the wind.

Moving as stealthily as I could, I laboriously approached the sett where paddock met copse above a small stream. The trickle of water was lost in the breeze that mussed the branches. I stopped in the shadow of a mature oak. Two twigs cracked under foot, as loudly as a gun. Even in the dark, the enormous spoil heaps of lumpy clay clawed aside by the badgers showed up deep maroon. Their earthworks resembled a tumbledown medieval fort, in miniature. All around were shallow pits, latrines, neatly filled with pyramids of badger poo.

The wind whipped tears from my eyes and they rolled down my cheeks as I stood there silently, torch flicked off, watching blackness. If I had been suddenly illuminated by someone else's torch I would have looked like a disconsolate toddler. I had not gone roaming alone in the countryside at night since I was small, and childhood sensations suddenly washed over me. I had missed out on badgers as a boy and now, as an adult, reluctantly working in London, I missed out on spending time in the country. Searching for badgers, and the world they inhabited, was also my own idiosyncratic attempt to escape the strictures of suburbia and rediscover a state of being where I could be absorbed by the rush of the wind and the sway of the trees and the scent of the earth.

The thick round copse on the brow of the hill to the south-east was just as I had imagined the wood in *Danny, the Champion of the World*, the sort of place where poachers go in search of pheasants. Although I felt a genuine frisson of fear, an alertness that comes from thwarted senses, it was a small revelation to me that I was no longer afraid of the dark. Some kind of toughening up, or disenchantment, had taken place since I read Roald Dahl's story as a boy. With badgers on my mind, I actually identified not with Danny but with the gamekeeper, the shadowy figure standing stock still in the lee of a thick tree trunk. The watcher, not the watched.

A dead branch reached down to brush my shoulder like a hand. I jumped. The wind met the bare branches, and together they roared like waves, blotting out all else for twenty seconds. Then the wind subsided for a minute, before it raced through the trees again. A tawny owl called, a low note, in the distance. Another one, higher-pitched, answered. On the nearest rise, a field away, a cluster of trees rubbed their branches together with a squeak that sounded like a sudden twist of the dial when you tune an old-fashioned radio. Then it became more exotic, a flamboyant scrape that could have been the complaint of a tropical bird. I was satisfied with my own silence until the wind dropped. Suddenly, everything about me was audible, and I felt completely exposed. My jacket rustled, my stomach gurgled and my right knee joint creaked. A bad-tempered sheep sounded as if it was single-handedly rounding up its flock, marshalling them against the menace of the night. The stream reverberated in stereo. Distant cars swept by on the A39 to Bath. A barking dog, a mile beyond.

In this cacophony, nothing rose up from the earth. No black-and-white striped masks staring hard at me, no trundling humping blundering grunting silhouettes; no sniffs of the air, no snorts, barks or groans; no musky smells. No badgers. My thudding boots and human stench probably scared them off. Or perhaps it was too cold, and the badger babies were too tiny and mewling to leave alone in the fug of the sett without a warm, lightly furred belly curled protectively around them.

Feeling thwarted but oddly enlivened by standing alone at night in a random field, I wandered to the crest of the nearest hill, where a few trees clung. There was a flash close by, and something pale rushed against my vision. An owl. A security light by a cottage on the eastern horizon flashed on and off, as if it were Morse code, and then it was snuffed out, quickly, by a shadow passing across it. This spooked me. For a second, I saw a man running, in the dark across the field. Then I realised it was a mist, tumbling down the valley but travelling with purpose, as if it were a wraith. My imagination was taking over, filling in the blanks of the night.

I stuck closely to the shadow of the hedges, still vaguely hoping to chance upon a badger on its nightly rounds. Nothing was abroad. The grass was wet, the earth cold, and I now understood what chilled-to-the-bone felt like. I sought out the light of the farmhouse like a moth and, once inside, it took a few minutes in the warm before my fingers began tingling hot–cold, deliciously, as the cat Moz ('Mask of Zorro') curled in a perfect comma on the floor beside the wood-burning stove. It was going to be harder to find the inhabitants of Badgerlands than I first thought.

2

Meles meles

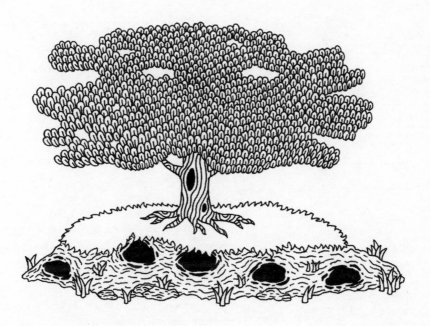

A dome of dark-leaved trees smothering a broad hill, Wytham Woods appears to hover, like a fecund flying saucer, over the floodplain of the Thames. These 1,000 acres of ash, maple and hazel are the epicentre of Badgerland. Here, four miles west of Oxford, lives a higher density of badgers than anywhere in the world. But although the wood on the hill may be perfect for badgers, when I arrived there one early summer evening it seemed that humans were not so welcome. Wytham's owners were determined to keep hoi polloi at bay. Unsignposted, the wood was reached via a road marked 'private – no thoroughfare'. At the gate there were further instructions – members of the public could visit but must first apply for a permit. Once granted, by the 'conservator' of the property, the visitor must stick to the footpaths, depart before dusk and obey a long list of bylaws, which prohibited fun: nothing must be collected or removed from the wood and there must be no dogs or horses, no bikes, no fires, no metal detectors and, heaven forbid, 'no playing of radios, tape recorders and musical instruments'.

Most importantly of all, the scientific equipment must not be touched.

This sobriety was because Wytham was dedicated to science. Since it was bequeathed to Oxford University during the Second World War, the wood had become the most renowned natural laboratory in the country. It was one of the birthplaces of the science of ecology and the site of many significant scientific undertakings, most notably a famous study of great tits which began in 1947 and, still going strong, is thought to be the world's longest-running population study in which wild animals are ringed or marked. These four square miles were also home to at least 220 of the best-studied badgers in the world. For the past five decades, the badgers of Wytham have been trapped humanely and weighed, measured and scrutinised by curious scientists, beginning with the pioneering work of Hans Kruuk in the 1970s and continuing today with a twenty-five-year study overseen by Professor David Macdonald and his project managers Dr Chris Newman and Dr Christina Buesching.

Barred from exploring this badger metropolis alone, I waited by the White Hart in the village of Wytham (pronounced 'Whiteham') for Drs Chris and Christina, who had agreed to take me beyond the barricades to catch a glimpse of how we acquired our scientific understanding of the badger – its habits, its homes, and how it socialises, communicates, lives and dies. As smoke from a summer barbecue wafted over the pub wall, Chris and Christina came clattering around the corner in an exhausted university-issue Land Rover. A couple who met in Wytham while examining badgers as students,

Chris and Christina were wry global intellectuals who commuted to Oxfordshire from Nova Scotia and collectively possessed the biggest brain ever to be trained upon these animals.

The badger's common names reflect our first impressions: its black-and-white colouring and its appetite for digging. The name 'badger' could simply come from 'badge', a description of its black-and-white striped face, and may also be derived from *bêcheur*, a French name for the animal meaning 'digger' introduced to Britain by the Normans, who also brought new breeds of dog across the Channel to hunt badgers. The modern French word for badger, *blaireau*, is from the Old French *bler*, denoting 'striped with white'; *blaireau* also means 'corn hoarder'. Almost as frequently used in Britain is 'brock', a Gaelic-Danish word for badger, while 'pate' was once the favoured name in the north. The badger has also been called a 'grey' and 'bawson', a term that originated in Scandinavia and means striped with white.

The badger's scientific classification as a mustelid points to its less obvious distinguishing features. A family of carnivorous mammals, the Mustelidae are a varied bunch: creatures of water, trees and earth. They tend to be long-bodied, sinuous and predatory – weasels, stoats, pine martens and otters – and so, on first impressions, the badger looks like the black sheep of the family. As Tim Roper, emeritus professor at the University of Sussex and author of *Badger*, a thorough and very readable tour of what we currently know about *Meles meles*, puts it, an animal can survive by being quick and agile or robust and strong; it can be a sniper or a tank. And the badger is a tank. When faced by an adversary, it adopts a 'head down' stance, displaying its stripes as a warning. 'If it is meant to be intimidating, it works,' he writes. But the badger also has a deceptive sinuosity more in keeping with other mustelids, alongside their sharp teeth and long front claws.

All carnivores have anal glands but the badger has a special, large scent gland, the subcaudal gland, beneath its tail. The badger's habit of scenting, or musking, baffled early badger watchers. 'There is no doubt the badger sucks and licks this substance, whether by way of taking a tonic, a cooling draught, a stimulant, or other physic, I cannot say,' wrote Sir Alfred Pease in 1898. But rather than the slurping of an energy drink, scientists discovered, the badger's rather philosophical squatting on its nightly rounds is actually an eloquent expression of identity and territory. Badgers use their distinctive musk to recognise each other, signify their reproductive state, whether they are a baby or an adult and, crucially – where badgers live in clan-like social groups – whether an individual animal is an insider or an outsider.

Twenty million years ago, the ancestors of *Meles meles*, the European badger, began evolving from their marten-like origins as agile climbers with long tails and a predatory demeanour. To survive in the evolving grasslands and savannahs, prey animals could either run, as ungulates did, or burrow into the ground like rodents. Predators had to be either good at chasing, such as dogs, or good at snaring subterranean prey, in which case they evolved to slink down holes like a weasel or became rotavators, such as bears and badgers, with formidable claws to dig for their dinner. In the badger's case, they could also burrow away from larger predators. Around three million years ago, the ancestors of the badger moved north into Europe. The earliest fossil remains, of Thoral's badger, *Meles thorali*, a clear precursor of today's badger, are about two million years old and were found at Saint-Vallier, near Lyons. At Boxgrove, West Sussex, badger bones dating from the Pleistocene era, between 750,000 and 500,000 years ago, have been found alongside the bones of human ancestors, which suggests the British relationship with badgers is an ancient one. The last glacial period, between 110,000 and 12,000 years ago, forced the badger to retreat from much of northern Europe until the thawing of the ice enabled it to colonise Britain in the company of bears, wolves, Arctic foxes, reindeer – and humans.

Wytham is a perfect illustration of how badgers across Britain have successfully adapted to a landscape increasingly shaped by man. When the ice retreated and badgers advanced again, they would have found Oxfordshire to be inhospitable Arctic tundra. As the climate warmed, forests of birch and pine quickly established themselves, followed by dense deciduous woodlands of oak, elm, lime and hazel. The Thames Valley may have been a swampy floodplain, home to beavers and too wet for badgers, but the woods of Wytham were high and dry. An outlying piece of Cotswold escarpment, the hill possessed a limestone top, below which was a generous layer of sandy soil that met the heavy Thames clay at the bottom. While Neolithic people created modest clearings across Oxfordshire, badgers dug their setts in the easily mined sand halfway down the hill at Wytham; the limestone cap above their heads was a perfect roof and stopped their tunnels caving in. The distribution of badgers in Britain today remains influenced by geology; badgers can't dig setts in heavy clay soils that will collapse or flood and so are largely absent from valley bottoms, floodplains and swathes of flat land such as East Anglia.

No slice of British countryside endures without a reason, and Wytham Woods escaped destruction because of the same quirk of geology – the limestone cap that made it so ideal for the badger. Half the forests that covered Oxfordshire had disappeared by Roman times; by 1300 just 7 per cent was left. Like everywhere else, Wytham's ancient trees were taken for firewood. The University of Oxford joined in the destruction – in 1632 eight oaks were purchased for just over £11 to make the grand gates of the Bodleian Library's Schools Quadrangle – and the trees on the hill were divided by a toll road known as the Singing Way because medieval monks on pilgrimage from Cirencester to Canterbury broke into song when they caught sight of Oxford for the first time from the high ground. But the woods endured.

In the Middle Ages Wytham was earmarked for tithed agriculture, but when farmers tested the soil they realised that the limestone below would break every plough. After Henry VIII seized control of Wytham from Abingdon Abbey during the dissolution of the monasteries, he sold the woods to acquisitive aristocrats who planted new forests for hunting. Assisted by the Enclosure Acts, the 5th Earl of Abingdon carved a much larger estate from common land and liberally adorned it with beech, oak, lime, sycamore and wych elm. The family's fortunes changed, however, and in 1920 the penniless 7th Earl sold his 3,000 acres to an arriviste miner called Raymond ffennell. The London-born son of a German father and English mother, Raymond Schumacher, as he was originally known, was educated at Harrow before making his fortune in the gold mines of Johannesburg. In 1915 he returned to England, adopted his mother's maiden name to avoid anti-German sentiment, and bought the estate of Wytham, including the woods. At first, ffennell, his wife and daughter Hazel commuted from London and stayed in 'the Camp', a set of lavishly carpeted, heated and fully staffed tents erected on the hilltop. Then he built a hunting lodge in the style of a Swiss chalet in the middle of the woods and, playing the perfect English gentleman, invited friends over for shooting weekends.

In the year the Second World War began, ffennell's heart was broken when Hazel, his only child, died of mysterious causes. She had been a sickly, rather solitary figure who had savoured the wildlife in Wytham, recording its beauty in her diary. With no heirs, and an inclination towards philanthropy – he had given his Johannesburg home over to sick children when he returned to London – ffennell bequeathed his entire estate to Oxford University. He wanted the woods to be known as the Woods of Hazel, and asked that their beauty be preserved for education and science, a bequest hailed in *The Times* as the most significant for the university since the Middle Ages. With the characteristic ruthlessness of a powerful landowner, the university initially ignored his wishes, signing an agreement with the Forestry Commission that dedicated the Woods of Hazel to the production of timber. Thankfully, the growing number of university fellows studying ecology recognised its value as a nature reserve rather than a commercial forest. The felling of ancient trees was halted in the 1960s and Wytham – the name Woods of Hazel never stuck – was truly given over to science as its benefactor had intended.

The geology of the woods was one reason for the density of the badger population at Wytham. Another was its bountiful supply of food. The fertile plains of the Thames Valley, which made the woods a virtual island, were hopeless terrain for setts but provided ideal foraging grounds. After dark, badgers would surge out of their elevated woodland home into the fields below to feed.

'The Badgerd battles much with sleepe, and is a very fat beast,' wrote George Turbervile, a sixteenth-century hunting enthusiast who was among the first to write of the badger's natural history. Badgers are great gastronomes. They enjoy eating anything we do – and plenty that we cannot stomach. In 1973, a Sunday paper published a photo of a badger that regularly visited a pub in Wales and was served a tipple of beer. As part of a series of 'repellent' experiments to discover what badgers found unpalatable, Chris and Christina once deposited some extremely spicy curry in the woods. They could tell it had been enthusiastically consumed by badgers when they found lurid yellow turmeric-coloured turds all over the place. Like us, badgers eat mushrooms, wheat, maize, oats, truffles, and have a bit of a sweet tooth; they enjoy plums and other fruit from trees, grapes from vineyards, and display a Pooh Bear-like gluttony for honey – the naturalist Denys Watkins-Pitchford, better known as the children's author BB, once watched a badger with 'the honey dropping down from each corner of his champing jaws in long elastic strings like yo-yo's'. Unlike most of us, they also love snacking on slugs, frogs, beetles, roots, bulbs, mice, voles, moles, birds' eggs, baby rabbits, rats, hedgehogs and the occasional fox cub.

Above all, though, badgers adore *Lumbricus terrestris*, the 'night-crawler' or lobworm, otherwise known as the common earthworm. An average badger can meet its energy requirements by scoffing 175 of these 10g worms in a night. My grandma once watched a sow badger devour 250 earthworms before midnight. 'She would snuffle about until she found some, lying as they do on the wet grass with only their tails anchored, and then with a tug and a stretch she would jerk out the worms and suck them down like spaghetti,' she wrote. This smart technique ensures the worms are not snapped as they are prised from the soil – because half a worm would be a terrible waste. In autumn, badgers eat to excess, building up stores of fat to survive a long winter.

Alongside its geology and plentiful food, one final element made Wytham a badger's idea of heaven: the protection of the woods and its devotion to science. Persecution has been a perennial part of the badger's life in Britain. But the strand of cruelty found in centuries of interaction between humans and badgers has been absent from Wytham's recent history. In fact, the worst treatment meted out to a badger in Wytham is probably an offensive epithet.

Until they relocated to Nova Scotia where they spent most of the year studying the impact of climate change on mice and voles from a log cabin called Ivory Towers, Chris and Christina lived alongside the badgers in the ffennells' strange Swiss chalet, a multi-tiered creation that resembled a wedding cake on the brink of collapse. In this slightly creepy dwelling set in one of the most crepuscular corners of the wood, their relationship with badgers became almost intolerably close. One evening, they returned home and wondered who was making such a noise in the downstairs bathroom. They opened the door, and there were three badgers, queuing up to drink from the toilet bowl. One badger they named Little Bastard on account of his predilection for nocturnal vandalism. On another occasion, woken by a terrible clamour in the night, Chris had to free Little Bastard's head which, like a cartoon character's, had become stuck after it plunged through a wooden door. 'There were some new words I learned that night,' said Christina, who was born in Germany.

We parked the Land Rover beside the funny old chalet and Chris and Christina ushered me into a large gloomy room at the rear that smelt one part chemistry lab and one part stable. The badgers of Wytham had been meticulously monitored since 1987 and this was the badger-processing HQ, where the scientists conducted their annual three-week census. Patients from a local drug rehab centre helped them lug traps, and badgers, around the wood; tough but oddly therapeutic work for the volunteers. Tomorrow morning they expected to catch around fifteen badgers; seventy individuals had already passed through here in the last week.

Once caught alive in the eighty-five sturdy wire-mesh traps dotted around the wood, each badger was sedated and inspected. First, its inside hind leg was examined. If there was no black ink tattoo already, the badger was given one. Unlike badger lovers, who name their objects of study, Wytham's scientists use numbers. The previous night Chris and Christina had reached badger no. 1,436. Tattooing sounds old-fashioned but it was quicker and more reliable than using microchips. As well as getting a tattoo, each badger had a distinctive symbol clipped into its guard hairs. This helped the researchers identify specific animals on infrared video surveillance, crucial for behavioural studies.

Each individual had its own record sheet, noting its length, weight and general health, and the results of all manner of tests. DNA and blood samples were taken and parasites and ectoparasites living in and on the badgers were logged. A study of badgers in East Sussex found 88 per cent had fleas, but more deadly than these ectoparasites was a gut parasite, coccidia, which, according to Chris, who studied parasitology, often determined whether a cub lived or died. If a cub survived into adulthood, it gained immunity from it. In a good year, Wytham's badger population rose above 300. In an average year, fifty cubs were weaned and made it above ground (those that die in the setts are never counted). Badgers may be robust and Wytham safer than most other places but mortality at birth was still high; in some years, more than 80 per cent of cubs died. In 2011, just fourteen were counted in Wytham: a hot, dry spring reduced the availability of worms and caused most of the cubs to starve to death in the first few sett-bound weeks of their lives.

The badgers had their hormones measured as well as the level of antioxidants in their bodies. Chris and Christina also peered into their mouths, where they saw some catastrophic dentistry. Although tooth-less old boars survived because they could still suck up worms, they often carried nasty bite wounds because they were less able to defend themselves in fights, a small sign that *Meles meles* is a more formidable adversary than many of us imagined. One final piece of data was extracted: every trapped badger had the skin pouch of its subcaudal gland gently scooped out with a small silver spatula and each individual's distinctive musk was deposited in vials in the freezer.

Christina is the world's leading expert in badgers' olfactory com-munication and one of her PhD students was currently conducting a scent playback experiment at Wytham, depositing 'foreign' badger scents in particular spots outside their territory and videoing resi-dents' responses. A resident badger would sniff an alien scent particularly intently and then spend considerable time searching for this phantom stranger. An individual lacking the scent of the local clan might be viciously dispatched.

If this smell was the defining characteristic of a badger, and one of the most important ways in which it expressed itself, I had to experi-ence it. Christina opened the freezer door and defrosted a small glass vial by warming it in her hands. 'It looks very much like mayonnaise,' she said. It smelt, added Chris, arching an eyebrow, rather like another popular household spread. 'If you ever wondered where Marmite came from ...' Steeling myself, I unscrewed the cap and inhaled one badger's special mayonnaise. It did indeed smell like extremely strong Marmite mixed with billy goat and an upper note of rancid fox. Unforgettable.

*

We left the laboratory and strolled through Wytham. Like someone at first turned away from a private club and then later admitted, my offence at the fussy restrictions curtailing access melted away once I was inside the woods. Our solitary walk at dusk was luxuriously exclusive. This elevated sanctuary felt as if it was suspended above Oxford. To the south, two hot-air balloons hung in the sky near Didcot power station. Remote sounds drifted up from the busy Thames Valley, amplified by the vivid acoustics of a domed, wooded hill. When they used to live here permanently, Chris and Christina would watch badgers to the blare of every barge party and summer ball. When an oil storage terminal near Hemel Hempstead, thirty-five miles distant, dramatically exploded in 2005, the noise arrived at Wytham with such clarity and violence that the couple thought their home was being broken into.

It looked as if the scientists, as well as the badgers, were jostling for space in Wytham. Trees had strange identity tags on their trunks, measuring growth rates, or modest square carbuncles – some of the 1,000 great-tit boxes and 500 blue-tit boxes in the wood. A metal disc on a pole gleamed in the gloaming, another mysterious measuring contraption. 'It's easy to bump into other research projects,' admitted Chris. On one occasion, he came across a group firing three 12-bore shotguns into an ash tree. This, apparently, was the best way to measure its seed production – the shot caused the ash keys to scatter onto blankets spread out below.

A barn owl swung low over a meadow and we passed an old gamekeeper's hut, its wooden cladding peeling off like a dead tree sloughing its bark, as if Hansel and Gretel had grown up and left the fairy-tale building some years before. Ducking away from the path and wading through thick leaf mould, we halted at the top of a steep west-facing slope underneath a heavy canopy of mature trees. Dug into the slope at the point where the limestone cap gave way to sandy soil were fifty holes. This huge sett had been inhabited by badgers for at least 475 years; the scientists knew this because after the dissolution of the monasteries, a deed was produced for Henry VIII to sell the woods which identified land by its proximity to the 'great oak badger sett'.

The badger is fêted for its fortresses. The collective noun for a group of badgers is a 'cete', which probably comes from the Latin *coetus*, meaning 'assembly' or 'coming together'. It is now more commonly used to describe a badger's home. Lacking both speed and camouflage, the badger's primary means of defence from bears and wolves, and later from man, has been its labyrinthine subterranean residences. Rabbit warrens and fox earths are mean dwellings in comparison with the grandeur of a badger's sett. 'As an engineering feat the badger warren all but rivals the beaver city,' wrote Mortimer Batten, the author of an early natural history of the badger published in 1923. 'I have known a badger to achieve in a single night a feat of excavation which would have taxed the strength of a strong man armed with a pick and shovel.' Make that a JCB. When the government dispatched seven men to measure a badger sett in the 1970s they took eight days to get to the bottom of it, unearthing a 'typical' sett featuring sixteen entrances, fifty-seven chambers and a maze of tunnels nearly a third of a kilometre in length. The badgers had excavated twenty-five tonnes of soil to create it. Like a beaver's chain of lakes, a sett can reshape the landscape, shifting hedge lines and transforming the neighbourhood ecology, which becomes governed by nettle and elder, two plants able to tolerate the nitrogen-rich conditions created by latrine pits brimming with badger dung.

Despite its size, the great oak sett was currently inhabited by only ten badgers, an indication that an immense earthwork does not guarantee a large population. Badgers usually move between different bedrooms, rarely sleeping in one chamber for long. A Polish study found badgers used twenty-five different rooms on average over a year, spending only 2.8 consecutive days in each one. Scientists have posited competing theories to explain this, including the need to be close to food sources and to maintain group harmony; the most plausible is that regularly changing chambers helps control parasites in these potentially fetid lairs.

The badger's mania for spring-cleaning is another way to minimise parasites. 'The badger is far and away the cleanliest wild animal that we have – indeed, it is the only one of our burrowing beasts which seems to possess any idea of healthful sanitation,' declared Mortimer Batten. One of the most distinctive behaviours noted by badger watchers is the collecting or removing of cartloads of dry grass, leaves, bracken and even bluebells for their beds. A further piece of sensible housekeeping is their refusal to foul the setts; instead they use specially dug latrines, which are usually beside their regular footpaths around their homes – and another way to demarcate territory. Badgers have been observed carrying large carcasses into their tunnels for meals but few remains have been found inside, suggesting they clear their setts of old foodstuffs, unlike foxes, who litter their earths with stinking scraps.

Most ingeniously of all, badgers may even create their own damp courses and heating systems. The biologist Tim Roper observed three setts in East Sussex where chambers were lined with plastic bags and old fertiliser sacks, on top of which the residents had created more traditional nests of dry grass, moss and dead leaves. In another study, two captive badgers kept in an artificial concrete sett chopped up hay and straw into a pile, which began to ferment, reaching an internal temperature of 38°C. The badgers built a normal nest near by, and moved it closer or further away from their radiator as required, repeating this technique over two successive winters.

Before I arrived at Wytham, I assumed that there could not be much more to discover about such a well-studied mammal, just as there could hardly be any great scientific debate over the lives of badgers. I was wrong. Almost every known aspect of a badger's being, it seemed, stimulated more questions, and further research. With the increasing use of DNA to define species, taxonomists even dispute the number of species of badger, which are found in all continents except Antarctica and Australasia.

There are relatives of the European badger that indisputably belong in its subfamily, Melinae: the Asian hog badger is *Meles meles'* closest relative; the American badger, *Taxidea taxus*, a more distant cousin; and there are also four species of ferret badger, much smaller animals, which live in Asia and look like stoats. Distant relatives include the formidable African honey badger, *Mellivora capensis*, which isn't really a badger at all, and two species of stink badger, which, as their name suggests, are actually more like skunks.

Until recently, the Eurasian badger was regarded as a single species with three subspecies: the European, Asian and Japanese badgers. After recent work by geneticists, it is now regarded as three distinct animals: the European, Asian (*Meles leucurus*) and Japanese badger (*Meles anakuma*). In Japan, badgers have been found to live alone, more like other mustelids such as weasels, which are solitary hunters. In chilly northern Europe and the dry Mediterranean, badgers tend to reside in mixed-sex groups that defend a common territory, but often these comprise only three or four individuals. In Britain and Ireland, many badgers now live in bigger groups of five to eight. In hotspots such as Wytham, the groups are even larger – sometimes more than thirty individuals. The reasons for the increase in the size of badger societies are particularly intriguing, and divisive.

Hans Kruuk is a legendary figure in the animal sciences and the presiding spirit at Wytham. He arrived here in 1972, fresh from studying the hyenas of the Serengeti, and with his doctoral student David Macdonald, who later became Oxford University's first professor of wildlife conservation, set up the most comprehensive long-term study of the badger ever undertaken. Kruuk was struck by the puzzle that badgers lived in groups but appeared to derive little benefit from doing so. Unusually for animals that might be considered 'social carnivores', badgers in northern Europe do not display much communicative or cooperative behaviour: they forage alone and don't seem to unite to bring up offspring or defy enemies.

Seeking to discover what purpose their apparent sociability served in their evolution, Kruuk watched the Wytham badgers through a pair of infrared binoculars purchased with the Nobel Prize money awarded to the Dutch ethologist Niko Tinbergen, Kruuk's former tutor. Kruuk's time in Wytham led him to deduce that badgers represented a fairly 'primitive' or early stage of sociality, in which groups had formed but cooperation had not really begun. He described carnivore social groups in terms of the Scottish clan system. Eastern Scotland could support many clans in small territories because the fertile land supplied people with ample food, whereas the poorer soils of the north-west Highlands supported fewer people and so resulted in fewer clans spread across much larger tracts of land. Badgers arranged themselves in a similar way.

In answer to the question of why badgers would live together if they did not benefit from cooperation, Kruuk's explanation was ecological rather than behavioural: the composition of their groups depended on the distribution of their food – chiefly, the availability of earthworms. Because badgers required different types of terrain to locate earthworms in various weather conditions (for instance, boggy Thames Valley pastures could be mined for earthworms during a drought that rendered worms inaccessible on the dry woodland floor of Wytham), a territory set up by a pair of badgers was generally large, contained a mosaic of habitats and could support additional animals. As Chris Newman put it, if this large territory was abundant in food, it would be both impossible and a waste of energy to hold exclusive rights to it. 'It would be like trying to defend Tesco from all the other shoppers,' he said. 'There's enough food for everyone.'

Kruuk's theory was called the Resource Dispersion Hypothesis and has been developed further by Professor Macdonald, who applied it to other carnivores, such as lions, which were also once solitary but became communal. Chris Newman and Christina Buesching were firm believers in the value of Kruuk's theory but admitted it was disputed in some quarters. I travelled to the South Downs to meet Tim Roper, the scientist whose recent badger book was the first volume of the famous New Naturalist series to be devoted to a single species. A compact man with a beard and speckled grey hair, Roper grew up in the Cotswolds, and at Cambridge University he studied how animals learned. 'I wanted to find out what it was like to be a wild animal,' he said. Later Roper joined Hans Kruuk's research team in Scotland over a summer. 'He is one of the world's best field mammalogists. It was a wonderful training,' said Roper, whose book is suffused with admiration for Kruuk.

The pair fell out, however, over Kruuk's theory. 'Scientists like ideas,' explained Roper. 'A lot of scientific activity in and of itself is extremely boring – it's the idea that comes out of it that is exciting. We all tend to over-interpret our results because we want them to be important.' Kruuk, he thought, over-generalised about badger diet, focused too much on earthworms, and 'jumped into explaining these patterns of behaviour too early'.

Roper's nights watching badgers on the South Downs led him to question Kruuk's emphasis on the importance of earthworms in determining how they arranged their lives, and he argued that a badger's territorial scent-marking was motivated by mate defence rather than defence of food; it was rather like birdsong, which is all about protecting a partner. Kruuk's theory suggested that young badgers would stay in the home sett if there was enough food; Roper thought badgers would naturally disperse – leave home – if they could. He argued that the increasingly large groups of badgers in Britain were not created by plentiful food but through lack of dispersal opportunities. Badgers were trapped. Wytham, in fact, appeared to be vivid proof: so densely populated that there were few good vacant spots to dig new setts and establish new territories.

Roper's ideas about badgers were not revolutionary (comparable theories were generally accepted in the bird world) but he believed that 'the Oxford group', the badger watchers of Wytham, were still dogmatically fixated upon the idea that behaviour was determined by food supply. In their defence, Chris Newman pointed out that their genetic work showed that half of cubs born in one social group were fathered by males from another. 'So if groups are geared to mate defence then it is a rubbish strategy – and evolution doesn't favour ineffective strategies,' he said. Video evidence, he went on, showed that badgers did not defend mates, and without this mixing of mates between groups, inbreeding would weaken populations. Wytham's scientists maintained that Roper's argument could be incorporated within Kruuk's Resource Dispersion Hypothesis. Food was only one 'resource'; other resources shaping a badger's social group were the availability of mates, sett habitat and even bedding material. 'A parish needs a baker, a butcher, a church, a pub, a dairy, maidens, jobs,' as Chris Newman put it.

This seemed to make sense but I wondered if it left Kruuk's theory so broad that it lacked much predictive value. 'We tend to try and categorise every aspect of animal biology,' observed Chris Cheeseman, the biologist who established the legendary badger research station at Woodchester Park in the Cotswolds. 'This is not a wise or necessarily helpful approach, especially with species as adaptable as badgers. Nature is far more complex than our capacity to understand it.' I was beginning to see just how many badger mysteries were still to be solved. Had unusually large social groups coalesced in Britain because badgers had nowhere else to go? And did such large subterranean communities give rise to diseases such as bovine TB?

I had another question. Kruuk described badgers as 'primitively social' – did he mean that badgers are a bit stupid? 'I can't abide this term,' said Chris Newman. 'They are just as well adapted along a path that has suited their evolutionary trajectory as any other weasel.' Chris saw the British badger as the product of a farming system that served up such plentiful supplies of worms that a once solitary animal was now squeezed into high-density living. In these conditions, its fundamentally weasel-like mating system of 'mate with who you meet' created a promiscuous badger society. I was later to encounter plenty of badger enthusiasts who observed what they believed to be monogamous pairs, but the scientists of Wytham were having none of it. In fact, a study by Hannah Dugdale, a former student at Wytham, concluded that female badgers living at high densities allowed themselves to be regularly mounted by many different males to mask paternity and reduce the risk of infanticide, as well as cut back on male aggression. This, at least, sounded smart.

'So many of our ideas about intelligence are anthropomorphic,' pointed out Tim Roper on the question of the brains of a badger. 'Much of it has to do with social communication. Cats have little of that because they are solitary animals. Dogs have a lot of social communication because they interact with their owners.' We may anthropocentrically judge intelligence by how much animals interact with us, but animal behaviourists do actually view species' levels of social interaction as one broad gauge of it. Another is to look at an animal's way of life: active hunters need to be cannier than foragers or grazers, and so in taxonomic terms carnivores are seen as the next-smartest group after primates and monkeys. *Meles meles* may be carnivorous and live in groups, but given that they have a limited social life and are less predatory than other mustelids 'you would not expect badgers to be very intelligent,' judged Roper. Scientists agree that the badger's social life is limited. As Chris Newman said, they do not live together because they enjoy one another's company. 'They don't do meerkatty things.'

I was soon to learn that badger lovers cannot resist ascribing emotions and family lives to the badgers they watch; but scientists say that badgers display little altruism and don't, for instance, warn each other of danger as other mammals such as meerkats do.

Kruuk also called badgers 'inarticulate'. His successors at Wytham were slightly kinder, analysing badger sounds and identifying at least sixteen discrete noises, from high-pitched yelping to low-pitched bubbling and a bossy quack that resembles the call of a moorhen. One noise in particular has intrigued naturalists over the years: the badger's eerie wail. Eric Simms, a BBC natural history sound-recordist, who taped twelve distinct forms of badger chat in the 1950s, watched a boar and a sow come to an abrupt halt when they met one night. 'Half a minute went by, and then the still summer night was ripped apart by a most dreadful scream,' he remembered. 'Louder and louder it became; on and on it went. My scalp tingled while this terrifying scream went on without interruption for over four minutes. At last it stopped and a strange silence rolled over me like a swirling fog.'

This scream appeared in another mystery first mentioned by post-war badger watchers. One night in the 1940s, the naturalist Brian Vesey-Fitzgerald heard a sow who had apparently lost her mate make a 'weird unearthly cry' at the entrance to her sett. He watched, transfixed, as the badger excavated a large hole in a nearby rabbit warren. After a tense, apparently ritualistic, encounter with a boar badger, both animals disappeared. A short time later, the boar came back dragging a dead badger by a hind leg as the sow helped from behind. They reached the warren, placed the carcass in the hole and covered it with earth. Was this a funeral? Do badgers bury their dead?

The countryman Phil Drabble, who reared and wrote about badgers in the 1960s and 70s, recounted many instances of badgers 'walling in' dead comrades, sealing up the chambers where they had died. Another naturalist of that era, F. Howard Lancum, wrote of a Shropshire badger watcher who excavated a sealed-up hole in a sett and discovered the large body of an old badger, blocked off from the rest of the sett on the inside as well as the outside, which no human killer of a badger would bother to do. While scientists usually have little time for such anecdotes, Chris Newman said they often found exhumed skeletons of badgers at Wytham that had been walled into underground mausoleums and later excavated by their descendants. Tim Roper also noted that badgers had been spotted moving carcasses from the roadside.

These accounts may be another facet of the badger's pragmatic cleanliness and self-interested desire to minimise the parasites that quickly infest an underground home. And in human circles, the endurance of the idea of a badger 'funeral' speaks of many people's determination to believe there must be a social or even spiritual side to this stubbornly mysterious animal.

Back in Wytham it was almost dark, and time to leave Chris and Christina to continue their studies. 'While we were talking,' said Christina as we turned away from our position standing on the hill above the great oak sett, 'there was a badger down there looking up at us.' The animal had long slipped away, completely unnoticed by me, and I was amused that these scientists who worked so closely with badgers would not think to excitedly point one out – unable to conceive of a person for whom they were not an everyday occurrence.

Going out to watch badgers, Chris admitted, bored him senseless and he and Christina obviously didn't fancy holding my hand until I saw one. Despite failing to spy a single resident of the badger citadel of Wytham, there was something enchanting about the long dusk on the hill over Oxford. 'It's not a bad office really,' shrugged Chris as he clunked the Land Rover into gear and bounced us back towards the city. 'Another day, another badger' was their avowedly unromantic motto during the trapping season. But even these academics, who possessed the detached, slightly irreverent attitude of many scientists towards their subject of study, conceded there was something special about the badger. As Chris and Christina deposited me in the pub car park beyond the boundary of Wytham, I asked what they thought of badgers. 'If you were an animal, what would you be?' pondered Chris. 'I would be a badger – a stomach on legs, motivated mostly by food, that sleeps a lot.'

'And,' added Christina with a glint in her eye, 'nothing can get in your way.'

Nothing, she might have added, except man, and his dogs.

3

Foe

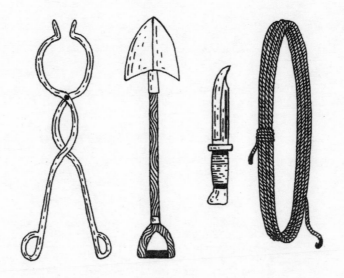

To do battle with such a formidable foe as the badger, it was advised in the sixteenth century, a man must find the following: a dozen strong men to dig; a dozen good dogs to work under ground and, for each, a collar with a bell attached; broad and narrow picks; a large spade; wood and iron shovels, a stout pair of long-handled tongs; sacks to stow the captured animals; a water bowl for the dogs; half a dozen rugs to lie on and listen, ear to the ground, for barking; Indian game fowls, hams and beef tongue to eat; copious flagons of alcoholic refreshment; and a little pavilion to light a fire for warmth in winter. 'Further, to do the thing properly,' wrote our badger hunter, 'the Seigneur must have his little carriage in which he will ride, with a young girl of sixteen to seventeen years of age, who will stroke his head while he is on the road.'

This randy, boozy, fowl-munching pederast who reclined in his carriage and lolled in his heated shed, in which he could 'donner un coup en robbe a la nymphe', while a small battalion of pickaxe-wielding labourers and rabid dogs waged war on a couple of badgers, was Count Jacques Du Fouilloux. He lived in Poitou and his book on the art of hunting was a runaway success in France. There were sixteen editions in the sixty years following its initial publication in 1561.

Count Jacques Du Fouilloux was not an aberration. Badgers were hunted with enthusiasm in France and the relationship between human and badger was an uncomplicated one: predator and prey. Du Fouilloux's instructions influenced badger hunting in Britain: our earliest surviving accounts of digging badgers from their setts and 'baiting' them by encouraging dogs to fight them to the death are copied, almost word for word, without acknowledgement, from the French hunter's book, although the requirement for a teenage companion was lost in translation.

Specially trained terriers were sent into setts, where they would corner a badger and hold the animal there, barking to signal to the men above where to start digging. The men dug their way to where the dog had trapped the badger under ground; at this point, the bait could begin. 'Having got him with the tongs, draw him, and put him in a sack,' instructed Du Fouilloux. 'Then take him to some court or walled-in garden, and let him go, putting the little bassets [terriers] after him, for when he is warmed up he will attack a man like a wild boar. And for this sport one must be booted, for several times they have taken a piece of my stocking, and with it flesh that was underneath.' Du Fouilloux advised putting the captured badger in an artificial sett and training dogs to attack it. Young dogs' confidence should be built up by allowing them to kill badger cubs in the spring.

'This done, take all the fox or badger cubs to the lodge and have their livers and blood fried with cheese and grease, to let them eat it, showing them the head of their game at the same time.'

The British writer George Turbervile, who reproduced Du Fouilloux's recommendations in *The Noble Art of Venerie or Hunting*, first published in 1576, endorsed Du Fouilloux's more brutal suggestions, including either cutting out the badger's teeth while alive or removing its entire lower jaw 'so that the beast can show the utmost fury without being able to do any harm'.

For centuries, badgers were part of a panoply of creatures who lived, and died, solely for our gratification. The masses took an uncomplicated pleasure in dogfights, cockfights, and the baiting of anything from a bear or an ape to an otter or a donkey and, in the sixteenth century, the badger's physical characteristics and way of life were of interest only in so far as they assisted the hunter in dispatching his quarry. In Turbervile's book, an illustration of a badger – classified under 'vermin' – looks more like a snarling Tasmanian tiger, hair bristling with aggression on its hefty, long-legged, boar-like body. Badgers were featured for their strength and – usually falsely – accused of crimes against poultry and livestock, and Turbervile was in no doubt about their vicious bravery. 'I have seene a well biting greyhounde, take a Badger and teare his guttes out of his belly, and yet the Badgerd hath fought styll, and would not yield to death,' he wrote. 'There biting is venomous, as the foxes is, but they make better defence for themselves, and fight more stoutly, and are much stronger.'

In one sense, though, sportsmen like Turbervile were early biologists, the first to study the nature of the badger. These animals may be fat, he recorded accurately enough, but they can outrun humans; their lower jaw cannot dislocate like other mammals', their skin is loose so it is hard to inflict injuries, and their skull has a ridge of bone running up the middle like a Mohican, the sagittal crest, which means it greets a fatal-looking blow to the head with a shrug, as if nothing has happened.

It was aristocratic huntsmen who wrote the early accounts of badger baiting, because they had the education to do so, but mostly badgers were a peasant beast and hunting them was a rough-and-ready village pursuit. Ordinary folk were egged on by Parliament, which passed a series of 'vermin acts' in the sixteenth century identifying certain species as pests and offering a bounty for their carcasses. The 1566 Act for the Preservation of Grayne (which was not repealed until 1863) designated rats and mice as vermin, but also species that ate almost no crops at all, such as badgers. The head of a dead badger won top price – twelve old pence, as much as a fox. Parish records in a Cumberland village logged the carcasses of thirty-six 'paytes' (a local name for the badger) between 1685 and 1750. The numbers aren't spectacular but, then again, the locals had another good use for badgers they caught alive.

Across northern Britain, many small towns such as Keswick and Kendal had bullrings, where bulls were baited with bulldogs and bull terriers. This widely accepted sport was held in public, unlike the comparable baiting of badgers, which occurred in private, in the back-rooms and courtyards of pubs.

The disreputable character of baiting was highlighted by the great Victorian writer-poacher Richard Jefferies. In 1879 he described the Wiltshire village of his childhood as 'a republic without even the semblance of a Government', venerating the ideals of liberty, equality and swearing. 'Betting, card-playing, ferret-breeding and dog-fancying, poaching and politics, are the occupations of the populace. A little illicit badger-baiting is varied by a little vicar-baiting.' A typical village pub bait was described by Mortimer Batten, who witnessed a badger caught by a gamekeeper and placed in a barrel in the stables behind a country pub near Whitby. It was kept in the barrel for three weeks, drinking a little water but eating nothing 'owing to the fact that it was constantly harassed and worried by its captors'. Every evening, the badger was placed in a box and terriers were 'tried' at him, tasked with dragging the animal from where it cowered. The badger would flash its jaws so fast that the dogs rarely succeeded. Mortimer Batten was invited to test his terrier against the badger, refused, and the next morning the badger was found dead in his box.

The most vivid account of baiting, however, was written in the 1830s in the poem 'Badger' by John Clare, the poet and farm labourer from Northamptonshire, who suggested that whole villages were complicit in the torture.

When midnight comes a host of dogs and men
Go out and track the badger to his den
And put a sack within the hole and lye

Till the old grunting badger passes bye
He comes and hears – they let the strongest loose
The old fox hears the noise and drops the goose
The poacher shoots and hurrys from the cry
And the old hare half wounded buzzes by
They get a forked stick to bear him down
And clapt the dogs and bore him to the town
And bait him all the day with many dogs
And laugh and shout and fright the scampering hogs

He runs along and bites at all he meets
They shout and hollo down the noisey streets
He turns about to face the loud uproar
And drives the rebels to their very doors
The frequent stone is hurled where ere they go
When badgers fight, then every ones a foe
The dogs are clapt and urged to join the fray
The badger turns and drives them all away
Though scarcely half as big dimute and small
He fights with dogs for hours and beats them all
The heavy mastiff savage in the fray
Lies down and licks his feet and turns away
The bull dog knows his match and waxes cold
The badger grins and never leaves his hold
He drives the crowd and follows at their heels
And bites them through the drunkard swears and reels

The frighted women take the boys away
The blackguard laughs and hurries on the fray
He trys to reach the woods a awkward race
But sticks and cudgels quickly stop the chase
He turns again and drives the noisey crowd
And beats the many dogs in noises loud
He drives away and beats them every one
And then they loose them all and set them on
He falls as dead and kicked by boys and men
Then starts and grins and drives the crowd again
Till kicked and torn and beaten out he lies
And leaves his hold and crackles groans and dies

Some critics believe the badger was a metaphor for Clare himself, hounded by village society, unable to support his family with his writing and alienated from his rural heartland. Others wonder if he wrote the poem as part of a campaign to get legal protection for the badger. Beaten to death and yet strangely unbowed, Clare's badger retained its integrity even as it succumbed to the tyranny of a whole community ranged against it. Baiting is easily condemned as evil and barbaric but perhaps there is excitement in Clare's portrayal as well, a thrill of the chase. Hunting the badger satisfied timeless human impulses. Daniel Defoe compared cockfights to 'the very model of an amphitheatre of the ancients'; and the contemporary literary critic David Perkins argues that baiting was an attractive spectacle because it encompassed suspense and the vicarious experience of fear, ferocity and courage. 'It was safely warlike, allowing spectators to indulge emotions of violence without danger to themselves. It satisfied whatever in the psyche makes it pleasurable for human beings to gang up on a victim.'

Perhaps the badger's great misfortune was that it never became a formally hunted game animal. The class-based hunting system controlled by the aristocracy that emerged in the eighteenth century imbued animals of the hunt with status and protection. Historically, red deer and foxes have depended on country sportsmen for their preservation in Britain. These species were conserved for the hunt, even more so after the enclosures enabled landowners to hold exclusive rights to game on their property. Sporting estates grew up, watched over by full-time gamekeepers. Clear rules governed how, and when, the game would be got.

Despite occasionally offering written instructions on how to kill a badger, the aristocracy never pursued the animal with much enthusiasm in Britain. The badger, it seemed, did not offer much in the way of diversion. It was not tasty like wild fowl and was never abroad in daylight hours, like the fox and deer. It could not easily be stalked, shot or hunted on horseback and, if pursued, ran quickly into the nearest sett. Digging it out was a back-breaking and tedious task ill suited to a gentleman. Baiting was disapproved of by most aristocrats because they considered it unsporting to stack the odds against the badger. But they had little love for an animal that served no purpose in human affairs and was too wild to be domesticated, and so offered it no protection.

The fact the aristocracy did not annex the badger for themselves probably made it more likely that farm workers and, later, working men in industrial towns would choose it as their sporting target. Badger baiting flourished as an affirmation of working-class community, a sport that labouring people kept for themselves and loved to practise in defiance of the ruling classes. This is not to stigmatise the rural underclass, for the landed gentry also persecuted the badger – through their gamekeepers – when it was judged to mildly interfere with their absolute dictatorship over the natural world. In other countries, where hunting remained a sport enjoyed by the common man, badger digging continued as an unproblematic country tradition. In Britain, any killing outside the rules of the hunt was defined as poaching. Too lowly to be targeted by the aristocracy but too common to be treasured, too slow to escape and too courageous not to put up a fight, badgers were caught in a class war.

It would be wrong to suggest that the badger in Britain was always treated with unremitting brutality over the centuries. Even before Enlightenment thinkers and Romantic poets firmly established notions of rights for animals, there were traces of a more nuanced conception of the animal.

The badger is not well represented in British folklore and was never regarded as cunning like a fox. Badgers tended to be viewed as old – as in the Yorkshire saying 'as grey as an old badger' – and secretive, or stubborn. People were 'as shy as a badger' or 'as hard to draw as a badger'; intransigent folk were 'as surly as a badger', which reflected how badly badgers responded to a life in captivity, tormented by dogs. Even the shortness of its tail inspired negative analogies: 'He's that greedy he would rob a badger of its tail.' The most important example of a more affectionate attitude towards the badger is an Anglo-Saxon riddle. It is written from a badger's perspective, an intrinsically sympathetic stance, and depicts the 'brave' and ill-fated life of an animal attacked by the 'death dog' and the 'stranger bringing doom of dying to my door'. The badger digs away from these hunters to save its young and retreats to a hilltop, before turning to attack its canine tormentor and 'viciously spear this creature I hate, who has harried me there'.

An emerging strand of concern for the badger became clearer towards the end of the eighteenth century. In 1790, the splendidly titled *General History of Quadrupeds* described badger baiting with dogs as an 'inhuman diversion' that was 'chiefly confined to the indolent and the vicious, who take a cruel pleasure in seeing the harmless animal surrounded by its enemies'. Many gentlemen hunters believed the badger was badly treated. In *The Life of a Sportsman* Nimrod (real name Charles Apperley) had his fictional hero, Francis Raby, aged ten, bait badgers with his terriers before graduating to more acceptable country pastimes, such as riding his stagecoach extremely fast. Nimrod argued the badger, being 'neither a depredator nor a thief', was 'entitled to an exemption' from the list of game that sportsmen pursued. Writing of Britain's national sports in 1825, Henry Alken called the badger a 'most quiet and harmless animal' of 'invincible and endless courage' and bemoaned its selection, 'with the Bull, for the pious national English purpose of baiting'.

Thinkers who first established the Romantic notion that animals had rights assumed that poor people brutalised animals to compensate for their own powerlessness and social inferiority. Mary Wollstonecraft wrote in 1792 that those who were 'trodden under foot by the rich' take revenge on animals for 'the insults that they are obliged to bear from their superiors'. Eventually, Parliament heeded these growing sentiments. Bull, bear and badger baits were at first considered unsporting; in 1835, the Cruelty to Wild Animals Act made it illegal to torture these animals with dogs. The outlawing of bull and badger baits may have been less about animal welfare and more the ruling elite's instinct for self-preservation. As the literary critic David Perkins points out, after the French Revolution there was more aristocratic hostility towards baits because they attracted unruly crowds. These gatherings of the great unwashed were to be feared.

The 1835 law did not forbid baiting a captive badger, so it was perfectly possible to keep a 'tame' badger in a pub yard and set dogs on it. In any case, the law was openly flouted. Tom Speedy, an avid Edwardian sportsman, found that long after it was banned, baiting 'was now and then indulged in by the lower stratum of society in rural areas'. He recalled as a child watching a baiting session carried out by Gypsies. 'I remember money changed hands and the language used, as is generally the case in such low species of gambling, was more expulsive than refined,' he wrote primly.

Despite this, the Cruelty to Wild Animals Act marked a significant change in our attitudes to badgers. From this moment, baiting was no longer defended by those who were respectable enough to write about it. Another form of hunting the badger, however, continued to thrive. From the nineteenth to the middle of the twentieth century, sporting authors drew a distinction between badger baiting and badger digging. The former may have been the preserve of uncouth pub yards but the latter, some country gentlemen insisted, was a reputable sport. Digging began like a bait, with a badger being located in its sett by specially trained terriers, who would corner it under ground until a group of men could dig down. Unlike a bait, the entertainment ended when the badger was caught and bagged; it would not be tested against dogs but was either swiftly killed or released. 'Properly conducted, badger-digging is no more cruel a sport than shooting or ratting,' argued H. H. King in a slim book that made a stout defence of digging in 1931. Badger 'hunting is legal, and the animal has every chance of escape. If he has not, it is not sport,' argued another enthusiastic digger, Captain Jocelyn Lucas, in the same year. By this definition, a bait was cruel; but digging, which might not end with the capture of a badger, was not.

In the nineteenth century, digging clubs dedicated to hunting the badger proliferated because of the Victorians' ardent passion for another animal – the dog. In 1833, four years before she became queen, Victoria wrote in her diary: 'I dressed dear, sweet, little Dash for the second time after dinner in a scarlet jacket and trousers.' Dash was her beloved dachshund, a hardy, short-legged dog bred by German and Austrian aristocrats to go under ground in pursuit of the badger (*Dachs* is German for badger). A few years earlier, the teenage John 'Jack' Russell was studying at Oxford when he encountered a milkman with a young bitch terrier he particularly admired. Russell bought the dog, Trump, and started breeding a distinctive line of white-haired and fiendishly tenacious little dogs. The Jack Russell terrier was born.

The Victorians worked hard to improve how society treated dogs. The banning of baiting in 1835 was primarily motivated by a desire to stop cruelty towards the dogs, which were maimed and killed in their encounters with bulls, bears and badgers. The Society for the Protection of Animals became a Royal Society with Queen Victoria's patronage; the Home for Lost Dogs was opened in Battersea in the 1860s; the Kennel Club was established in the 1870s to regulate breeds and breeders; and there was an explosion of dog shows, including the first Crufts. Many of these new animal lovers wanted to 'work' their dogs, and use them to hunt wild animals.

John Russell became Reverend John Russell but 'did not over-exert himself as a parish priest', according to his biographer, Eleanor Kerr. Instead, he took a sleepy tenure in North Devon and devoted his time to hunting – foxes, otters (he was fêted for killing 'the almost incredible number of 25 otters in the last two summers, for which he should receive the thanks of the fish', as Nimrod recorded) and badgers. He kept a pack of otter hounds, obsessed over his Jack Russells and became disillusioned with the new dog show scene's veneration of beautiful, insipid breeds who never did a day's 'work' in their life. 'Gameness' was the characteristic he valued above all else in a dog; a willingness to work, a yapping desire to confront wild animals bigger than itself in the service of its master. And game Jack Russell terriers were the perfect breed to confront a badger in its sett.

After the parson's death aged eighty-eight, Arthur Heinemann became the most important Jack Russell breeder. An Eton- and Trinity-educated sporting journalist, he bred animals directly descended from Parson Russell's line and established a badger-digging club in 1894. The objectives of the Devon and Somerset Badger Club, later renamed the Parson Jack Russell Club, were to promote badger digging and breed working terriers. 'Next to rat hunting, badger hunting is the best sport with terriers,' wrote J. C. Bristow-Noble in 1929. These hunters were not driven to breed killer dogs by some visceral hatred for the badger; rather, their passion for terriers led them to view the badger as a diversion and the ultimate test for their eager young pups. Digging clubs were formed all over Britain. The Kent and Sussex Badger Club even provided their precious terriers with badger-skin collars because only this tough skin would protect their throats from dog – and badger – bites. Captain Lucas did not approve. 'I always felt that it was a little theatrical,' he wrote.

One iteration of badger hunting was to pursue them at night. In 1948, the *Somerset County Herald* recorded a nocturnal badger hunt at Thornfalcon with an 'excited' field of more than 150 participants. Night-time hunts were known as 'giving Brocky a birthday'. Badgers would be watched coming out of their sett on a fine moonlit night and allowed to wander off for half an hour before hounds were set on their trail. 'The field follow as best they can, tumbling over banks, and falling into ditches or bramble bushes, but all enjoying themselves hugely,' wrote Lucas. 'Matters are simplified if each person carries an electric torch.' A few years earlier, the naturalist J. Fairfax Blakeborough recommended using policemen's lanterns and believed moonlit hunting could save the badger. 'If brock paid for his right to live in this way he would soon have something of the sacred halo around his head which surrounds that of the fox,' he argued.

Reputable diggers insisted most captured badgers were released, and sometimes even relocated their prize catches so badgers could repopulate regions from where they had disappeared. 'Badger clubs rarely kill the badgers, unless required to do so by the Hunt, or the landowners concerned,' claimed Lucas. As he suggests, some badger digging was a direct corollary of fox hunting: 'terriermen' employed by hunts to flush out foxes that had gone to ground would manage earths and setts for their local hunt. If badgers made their home in an artificial fox earth, the terrierman would dig out and kill the badgers so that foxes could breed there. Badger setts might also be blocked up if they were providing too much sanctuary for the fox from the hunt. In this way, the badger was collateral damage in the British passion for hunting foxes.

Badger digging, unlike baiting, was sufficiently respectable to be enjoyed by members of the ruling classes for much of the twentieth century. It was a Sunday activity, carried out after church.

An issue of *The Field* of 1921 included an account of a dig in Lincolnshire organised by Lady Charles Bentinck in which five large badgers were taken in less than ninety minutes. Captain Lucas described another dig at Withcote Hall held by Sir Geoffrey Palmer, Bt, and in the 1930s went digging in Leicestershire with Sir Arthur Hazlerigg, the 2nd Baron Hazlerigg and the captain of Leicestershire

Cricket Club. 'Don't forget to take out some beer or cyder for the diggers,' instructed Lucas, implying that the gentlemen didn't actually pick up a spade. 'There is no more pleasant holiday than a day's badger digging in September with good dogs, willing diggers, and a reasonably easy sett. Take out plenty of lunch for your party and a waterproof rug to eat it on ... The freedom from cruelty in well-conducted digs is apparent from the number of ladies who come out regularly and get as excited as anyone.' As well as women, professional pillars of the community also took part. In 1920 Lucas and an archdeacon, who hunted badgers into his eighties, paused mid-dig at 11 a.m., and stood, bare headed, to observe the two minutes' silence on Armistice Day. In the silence, a young lad heard a dog barking far below him, and after they had paid their respects to the war dead, they dug down seven feet and captured two badgers. As late as the 1970s, friends of mine in Cornwall remember their family doctor going on badger digs.

In their defences of digging, Lucas and King both put forward various 'sporting' rules they felt should govern the practice. Lucas said badgers should not be dug between February and June, when cubs were around. Most countries where hunting badgers is still legal adhere to this principle of a 'close season'. Citing a contemporary French author who advised, like old Du Fouilloux, gagging a badger or removing its canine teeth so young terriers could attack it without being hurt, Lucas thundered: 'Neither of these would be tolerated for a moment in any English-speaking country, nor can they by any means of imagination be called sport.' But there were never any widely accepted rules governing digging and it could always segue into baiting if a badger was thrown to the dogs. While King advised that a badger should never be hunted with a pack of dogs, or for several days, and disapproved of tongs, Lucas praised 'our humane modern "tongs" which grip without hurting' and recounted epic three-day badger digs with ten of his precious Sealyham terriers let loose in a sett on a single occasion.

One man who took a different view of badger digging between the wars was Henry Williamson, who later found fame as the author of *Tarka the Otter*. One St Valentine's Day in the early 1920s, still recovering after being invalided home following a gas attack in the trenches, the disenchanted war veteran went for a walk overlooking the ocean near his home in North Devon. It was a wild, billowy day and everything spoke of death. The island of Lundy rose out of the mist on the sea. 'The distant headland was like a dead animal lying on its side, the green flanks sunken on the hidden frame,' he wrote. Then he bumped into the local badger-digging club: an innkeeper, a farmer and a couple of labourers, three small boys, a long-nosed man 'who reminded me, vaguely, of a badger', his wife and their daughter, a teenage schoolgirl with flaxen hair and a permanent smile. They brought picks and shovels, a large basket of sandwiches, an earthenware gallon jar of whisky and twelve terriers. The master of the club wore a grey bowler hat, a red waistcoat, a tweed coat and a white stock fastened by a pin made from a badger's penis bone. (This slender bone, the baculum, is about 10cm long and turned up at the end.)

Stopping outside the sett, the master looked for badger-pad marks and found a grey hair. He twiddled it between forefinger and thumb and felt it was flat – and therefore belonged to a badger. A terrier was slipped from its leash and sent down the hole. The master drew a copper horn from beneath his waistcoat and blew three faint toots. The terrier snarled from somewhere in the ground below and returned 'with red-frothed jaws'.

Men began digging, with spade, two-bill, bar, pick and mattock. A second terrier was encouraged into the hole. After an hour, it slunk out, bitten on shoulder and lower jaw. A third was sent down. Williamson watched this scene, and considered the promise of spring, the gorse in flower, a kestrel laying her eggs and the sow badger giving birth to cubs, to be stillborn. 'Swaling fires for the gorse, tongs and hunting knife for the badgers, gamekeeper's shot for the kestrel, loneliness for man, these are the ways of life. Yet endeavour goes on; all things aspire to the sun and the sky,' he wrote.

Williamson was served nearly half a pint of whisky by the innkeeper. 'Soon the badger dig took on a jovial aspect. It was a survival of one of the oldest sports in Britain, going back to the time of the Normans, when the terriers, the earth-dogs, came over in the wooden galleys.' The master told Williamson the sport of badger digging improved the strain of terriers. 'I asked him if it improved the strain of badgers,' wrote Williamson caustically.

Men took turns to dig; and the terriers took turns to hold the badgers in place under ground. One dog was called Mad Mullah. He was 'so fierce that whenever he saw the stuffed masks, or heads of badgers, on the walls of the Lower House bar, he snarled and leapt up at them,' noted Williamson. Gradually the men dug down to where they could see Mad Mullah's tail. The hole was enlarged, the terrier backed out and the onlookers 'crushed back in panic': emerging from the sett was 'a flat head shaped like a bear's'. It disappeared, and the terrier rushed after it.

The master knelt down, holding a pair of iron tongs. The handles were three feet long and created an iron collar when closed around the badger's neck. Untried dogs would be allowed to fight a badger held in these tongs; without that brutal implement the dogs would be torn to shreds. Next up, the sow, which was heavy with young but still lunged at Mad Mullah more swiftly than the terrier could jab back. Eventually, she was held by the tongs and dropped in a sack. Once inside, both badgers lay quiet.

The master and his huntsmen made 'bloodthirsty cries' and drank more whisky, while the terriers howled in the wind. After the pregnant badger was released, the boar was stunned by a blow to the nose with a spade and the master pressed a knife into its throat. Blood gushed from the wound and 'fouled some early daisies,' observed Williamson with precise disdain. He 'heard the first lark of the new year singing' as the master knelt on the ground and hacked at the neck of the dead badger. The severed head was presented to the schoolgirl, 'whose smile, I fancied, was only on her lips, not in her eyes', and the four pads were cut from the badger's feet. A small boy came forward to be blooded and blood was smeared on Williamson's brow and cheek as well, because it was the first death of a badger he had witnessed. 'I felt I had been false to myself, and yet another thought told me such feelings flourished only in nervous weakness,' he wrote. 'With the dried blood stiff on my temples I climbed the hill, cursing the satanic ways of men, yet knowing myself vile, for they had not known what they were doing, but I had betrayed an innocent; and all the tears – weak, whiskey tears – would not wash from my brow the blood of a little brother.'

The theory that the badger was a victim of Britain's class-based system of hunting appeals to me, and yet there may be something else that accounts for our blood-thirst: ignorance.

We regard a lack of feeling for nature as a peculiarly modern condition, a symptom of our suburban lifestyles, and yet our failure to understand the badger is timeless. The badger has been misunderstood, and feared, for hundreds of years. 'He hath very sharp Teeth and therefore is accounted a deep biting Beast,' wrote Nicholas Cox, a sporting author like Turbervile, in 1677. This was accurate enough, but Cox then endowed the badger with a truly astonishing feature: 'His Back is broad and his Legs are longer on the right side than the left, and therefore he runneth best when he gets on the side of an Hill.' Cox does not stop to wonder how the wonky-legged badger turned around and ran home again without overbalancing. This myth endured in sayings: the conspiratorial seventeenth-century Anglican priest Titus Oates was described as 'uneven as a badger', while 'badger-legged' was directed at people and objects with legs of unequal length.

In the sixteenth century, the badger was divided into two species: dog badgers and hog badgers. One had dog-like paws; the other pig-like cloven hooves. The dog badger was a villainous, wolfish demon, devouring lambs, chickens, fawns and carrion; the hog badger was a snuffling shy-and-retiring vegetarian that ate shoots and leaves. Even today, as the academic Angela Cassidy has noted, perceptions of the badger often cleave to these categories: the diseased, predatory, out-of-control bad badger, and the home-loving, family-orientated good badger.

Fear and superstition have long clung to these nocturnal animals, which were accorded considerable power in medieval witchcraft. Country people would say that if you heard a badger call followed by the cry of an owl it was a harbinger of your death. A badger crossing a path behind you was good luck; in front, bad luck. A bone or badger claw hung around the neck gave the wearer the power to keep secrets – appropriate, given the badger's shyness. In 1812, David Naitby, a North Yorkshire schoolmaster, wrote that the badger once played a crucial role in a form of divination, citing an outlandish brew containing skull fragments from a hanged man, the tongues of adders, graveyard worms, toad hearts, crab eyes, wild dove blood, great flitter-mouse (bat), mole blood, bullock blood and seven drops of brock blood, mixed seven years from the day on which the man was hanged. It may be significant that the blood of the badger was the only defined measure in this baffling potion.

Ignorance, and fear, continued throughout the twentieth century. Some sensational accounts were wilfully misleading. In 1911, an advert highlighted an attraction at Leyburn Fair in Wensleydale: 'An Extraordinary capture at Stainton near Darlington of a Monster

Badger which terrified the district with its destruction of sheep and cattle'. Rather contradicting its alleged fondness for raw beef and mutton, the badger was fed on cod heads, which it loved. Three years later, Blakeborough commented on country folk's ignorance of badgers, citing a farmer in Cleveland who found an 'otter' in a steel trap which was actually a badger. The *Shooting Times* of 1930 recorded a gunman believing he had killed a bear when it was a badger. In the 1940s, Ernest Neal wrote of the lack of awareness of badgers among rural people. Twenty years on, Michael Clark, another badger watcher, asked an old countryman if there were badgers near his home and was greeted with bewilderment. 'What's badgers?' the old man asked. When Sylvia Shepherd reared a tame badger in Cumbria in the 1960s, two neighbours thought it was an otter or a new breed of dog.

It is no different today. Most of us are not only blind to badgers that live a few hundred yards from our doors but are also ignorant of their lives and needs – and what they can and cannot be blamed for.

The centuries of bounties, baiting and digging took their toll, and by the middle of the nineteenth century badgers were scarce in many parts of Britain. 'This animal becomes rarer year after year,' noted Richard Jefferies in 1878. In 1882 it was declared that badgers no longer bred in Cumberland; they were reported to be nearly extinct in the Lake District at the turn of the twentieth century, and were considered rare in Kent in 1908.

The cause of this decline was not simply persecution by the lower classes. The badger, at most a minor irritant to the hunt for inconsiderately building setts in which a fox might hide away, was deemed vermin by the increasing number of gamekeepers employed to promote fox hunting and pheasant shooting on country estates. Badgers were accused of vulpicide and killing pheasants and grouse because they possessed flesh-tearing fangs; if they looked like a carnivore, gamekeepers concluded that they must have the tastes and table manners of a carnivore. Badgers were shot and trapped, punished alongside raptors, pine martens and every other predator.

As the population of gamekeepers rose, reaching a peak of 23,000 before the First World War, so the badger became rarer. Badger diggers blamed keepers for its disappearance. In 1929, J. C. Bristow-Noble feared gamekeepers would 'speedily put an end to their existence', and in the same year Tom Speedy attributed the vanishing badger to 'the common use of the steel trap'. Perhaps he need not have worried, because many badgers displayed a Houdini-like talent for escaping gin traps. They possessd an instinctive suspicion of iron, surmised badger digger H. H. King, and traps commonly caught only a few grey hairs. Countrymen believed badgers – perhaps as cunning as a fox, after all – either walked upside down on the ceiling of their tunnels when entering and exiting their setts, to foil a trap at the entrance, or curled themselves up and performed an elegant forward roll over the best-laid trap.

The badger may have been a Houdini but even escapologists need help. At the start of the twentieth century, the badger appeared destined for extinction – dug, disturbed and driven from entire counties.

In the space of barely four decades, however, it returned to parts of the countryside from where it had been absent for years. In the first half of the twentieth century, *Meles meles'* relationship with the human population of Britain was turned on its head. The badger was transformed from an object of fear, superstition and rural torture into a cuddly hero for children and a revered symbol of conservation for adults. More than any other, one man, and one fictional badger, were responsible.

4

Mr Badger

In 1908, Mole caught sight of Badger peering from a hedge. Badger trotted forward, grunted, 'H'm! Company', and disappeared again. 'Simply hates Society!' explained the Water Rat. With that modest and perfectly rendered cameo, the character who was to transform popular perceptions of the badger stepped onto the pages of *The Wind in the Willows.*

The Edwardian era witnessed the transformation of the society and landscape of the British Isles. As the working classes and women secured the vote, the countryside was revolutionised by agricultural depression, ribbon development, the railways and, most of all, the motor car. The badger, too, saw its place in the countryside and its relations with the human population radically changed, mostly by one writer. Beginning by telling silly animal stories to his troubled son, and then developing a whole world populated by Mole, Rat, Toad and Badger, Kenneth Grahame offered an entirely new portrait of a despised, feared and, at best, pitied wild animal. Grahame kindled a romance with the badger that endured throughout the twentieth century and changed relations between men, women and badgers for good.

Far more people have encountered Mr Badger than have ever seen a badger living in the wild and, like generations of twentieth-century children, my first sense of the badger came from Kenneth Grahame's story, and E. H. Shepard's illustrations. I read the Shepard-illustrated *Winnie-the-Pooh* first and although *The Wind in the Willows* featured similarly charming pictures, I sensed that it was taking me into a darker, more grown-up, and utterly engrossing alternative universe.

Like most readers, at first I identified with the timid Mole, who guides us through the gentle beginnings of this tale of wood and river-bank. It takes a while before Badger crashes into the story and yet his presence looms large long before, and after, his entrance. It is the urbane Rat (in reality, a water vole) who first tells the innocent Mole about Badger, who resides in the Wild Wood. 'He lives right in the heart of it; wouldn't live anywhere else, either, if you paid him to do it. Dear old Badger! Nobody interferes with *him*. They'd better not.' Mole clamours to meet this enigmatic, apparently powerful animal but is first introduced to the ridiculous but lovable Toad, a vain, childish *nouveau riche* type who lives in Toad Hall and develops a fascination with the motor car. Rat explains the futility of inviting Badger to dinner – 'Badger hates Society, and invitations, and dinner, and all that sort of thing' – and it is only when Mole sets out on an irresponsible mission to explore the Wild Wood that he and Ratty, lost and terrified, stumble upon a doormat in the snow, the gateway to the home of this intimidating character.

Readers cannot help but feel the terror of Mole and Ratty in the Wild Wood before Grahame invites us all to experience the balm of Badger's company, and his home. Having navigated a series of spooky passages, Mole and Rat encounter what the critic and biographer Humphrey Carpenter calls the very 'heart of Grahame's Arcadian dream'. I defy a child or an adult reader not to be entranced by this vision of Badger's kitchen:

The floor was well-worn red brick, and on the wide hearth burnt a fire of logs . . . Rows of spotless plates winked from the shelves of the dresser at the far end of the room, and from the rafters overhead hung hams, bundles of dried herbs, nets of onions, and baskets of eggs. It seemed a place where heroes could fitly feast after victory . . . or where two or three friends of simple tastes could sit about as they pleased and eat and smoke and talk in comfort and contentment. The ruddy brick floor smiled up at the smoky ceiling; the oaken settles, shiny with long wear, exchanged cheerful glances with each other; plates on the dresser grinned at pots on the shelf, and the merry firelight flickered and played over everything without distinction.

Carpenter thinks this underground dwelling is reminiscent of Beowulf's mead-halls. Badger's sett is also built on Roman ruins, a sly nod to the inspiration Grahame took from the classics and a symbol, perhaps, of the endurance of the natural world (and badgers) which swiftly recolonises human civilisation when it implodes. Most of all, though, this description is a child's fantasy of warmth and safety, a 'womb-like refuge' according to Carpenter. Grahame's ideal kitchen may belong very much to his generation, where children of his class would linger in kitchens bustling with domestic servants, but it has a timeless appeal. It is not simply a fiction, either, for Grahame is faithful to wild badgers by making Badger's home such a defining part of our experience of him.

Mole and Ratty's arrival at Badger's sett is pivotal in *The Wind in the Willows*: once the lesser animals are cocooned here, Grahame's tale stops being a pastoral adventure story with talking animals and becomes more allegorical. Toad is terrorising the neighbourhood with his new obsession for cars, and Badger decides he must be 'taken in hand'. Toad defies Badger's natural authority, and that of the law, and despite being sentenced to twenty years for joyriding, escapes prison by dressing up as a washerwoman. When Toad returns to the riverbank he finds his beloved mansion has been occupied by an uprising of stoats and weasels, and it is only when Badger takes up the cudgels, with the help of Rat and Mole, that the hall is retaken from the lawless rabble, and peace is restored. Without Badger, Grahame seemed to be saying, this bucolic idyll, his fantastic version of the English countryside, would be lost for ever.

Kenneth Grahame was an unlikely revolutionary in Badgerland. The son of an alcoholic Scottish lawyer, he was a deeply conservative pillar of the establishment, who shared the anxieties of most upper-middle-class men of his generation: half-contemptuous, half-terrified of the rising power of the unwashed masses, dismayed at the ugliness of city life and profoundly uncomfortable with women and the sexual revolution. In his desire to escape these very Edwardian miseries, and his unhappy home life after a disastrous late marriage, he sought comfort in a fantasy of rural life populated by talking animals.

If everyone has an adjective, observed Grahame's contemporary C. L. Hinds, then *The Wind in the Willows*'s author was 'startled'. Grahame's mother died in 1863, when he was four. His father, James Cunningham Grahame, proved incapable of bringing up four children, and Grahame and his siblings were sent to live with Grahame's maternal grandmother in a rambling house close to the Thames in Cookham Dean, Berkshire. This tranquil existence was interrupted when Grahame, aged seven, and his two brothers and sister were transported back to their father's. Once again, Cunningham Grahame could not cope and fled to France, where he drank heavily and had almost no further contact with his children; Grahame was dispatched to boarding school. Already, his biographer Alison Prince contends, Grahame had 'entered into a conviction that the world of adult human beings was a treacherous and unpleasant place, and he never quite recovered from it'.

Throughout his adulthood, Grahame wrote unusually vividly about what it was like to be a small boy. Dreamy and poetic, he was desperate to go to Oxford University, but his academic aspirations were frustrated when he was installed as a clerk in the Bank of England by several uncles who exerted a controlling influence over his life. Despite his resistance to the life of a bank clerk, Grahame dutifully climbed the career ladder and, in his spare time, wrote essays. In 1895 he published *The Golden Age*, a childhood memoir which drew on his lost rural paradise and his unusually acute sense of what it was like to be seven years old. It received rapturous reviews and a book mined from the same vein, *Dream Days*, came out in 1898, the year Grahame was made Secretary of the Bank of England. Then, silence. With no childhood reminiscences left to mine, Grahame was consumed by his job and a decision, late in life, to become embroiled with Elspeth Thomson.

Reading of Grahame's life, I felt sorry for this sensitive loner, an emotional orphan who sought solace in childhood and the countryside, but it became harder to feel sympathy for him after 1899, when the forty-year-old bachelor married Elspeth, a wealthy, rather badly educated daughter of a Scottish inventor who had already formed friendships with Alfred Tennyson, the poet, and other writers before she met Grahame. Most biographers and critics of Grahame take the old-fashioned view that this naive gentleman was ensnared by a scheming groupie, despite Grahame being a more than willing colluder in the fantasy world they created. His letters to 'Minkie' were rich in baby-talk and what Alison Prince calls a 'competitive invalidism': the couple appeared to encourage each other to become ever more whimsical and incapable of coping with the adult world. Prince believes that Grahame could 'envisage no relationship except that which links child to child or child to authority-figure' and this appeared to be borne out in his treatment of his only child, Alastair, who was born prematurely with a squint in his left eye and his right eye blind from a congenital cataract.

Grahame was determined to give his son, nicknamed Mouse, the benefit of a father who was nothing like the remote 'Olympian' adults he disparaged in *The Golden Age* and *Dream Days*, and so Mouse was indulged by parents who believed him to be a genius. When he was just three and a half, Mouse devised an attention-seeking game: he would lie down in the road in front of new-fangled motor cars and force them to screech to a halt. Grahame appeared not to intervene even when the child bullied young girls. 'Eee as a way of usin is *boots* on the form & edd of a defenceless female,' he wrote in typical fashion to Elspeth. When Grahame asked why Mouse had slapped a small girl, his son replied because he wanted to, '& I ad no argment set up gainst that,' he wrote. Grahame's 'detachment from the moral judgement which he might have considered "Olympian" was total,' judged Prince.

In a letter to his wife dating from around 1904, Grahame wrote of playing with Alastair: 'The ole of the time I ad ter [s]pin out mole [s]tories ... there was a [s]tory in which a mole, a beever a badjer & a water-rat was characters & I got them terribly mixed up as I went along but ee always stratened em out & remembered wich was wich.' Despite spoiling Mouse, Grahame and Elspeth were often absent, and during a typical summer the boy was dispatched to Littlehampton with his governess for seven weeks while his parents holidayed in Cornwall. Rather than respond directly to his son's plaintive appeals for a visit in his letters, Grahame replied with odd fairy tales, describing the adventures of Toad that later appear in *The Wind in the Willows*.

Grahame and Elspeth moved to Cookham Dean by the Thames, where he had been briefly happy as a child. There, he was visited by Constance Smedley, an ambitious young American writer and editor, who set out to persuade the frustrated author to end his barren run and publish a new book. Smedley inveigled herself into the family's bedtime routine, discovering that Grahame and his son shared 'an unending story, dealing with the adventures of the little animals whom they met on their river journeys'. Encouraging him to turn the tales into his next book, she wisely observed how 'Mouse's own tendency to exult in his exploits was gently satirised in Mr. Toad, a favourite character who gave the juvenile audience occasion for some slightly self-conscious laughter'. Biographers point to various inspirations for Toad's character but the main one, indisputably, was Grahame's own son.

From this unhappy domestic life sprang *The Wind in the Willows* and its cast of boyish animals. Long before it was published, Grahame wrote in an introduction to *Aesop's Fables* of animals being 'exploited' in an 'ungentlemanly manner' in allegories (his selection of a hundred of Aesop's tales featured foxes, lions, wolves, cats, lambs, goats, mice, kites, peacocks, a wasp, a tortoise and a leopard but not one badger), and dreamed of a wood where the animals told stories in which humans 'point the moral and adorn the tale'. Rather than burdening animals with human failings – making the peacock vain and so on – Grahame's introduction to Aesop hinted at how he would write about animals in a radically different way. The moment animals were truly studied, he wrote, 'they were seen to be so modest, so mutually helpful,

so entirely free from vanity, affectation, and fads; so tolerant, uncom-
plaining, and determined to make the best of everything; and, finally,
such adepts in the art of minding their own business, that it was evident
a self-respecting humanity would not stand the real truth for a
moment'.

Increasingly unhappy with his day job in the City, Grahame finally
resigned from the Bank of England in 1908 and, five months later,
The Wind in the Willows was published.

It was not an immediate hit. The reviews were, at best, perplexed
and could not work out whether it was a book for children or for
adults, and whether its characters were animals, boys or men. 'Instead
of writing about children for grown-up people, he has written about
animals for children,' wrote the young Arthur Ransome in a typically
lukewarm verdict. 'All the animals had a very stirring time, and but for
their peculiar shapes they would well pass for first-rate human boys,'
judged *The Nation*. Critics today concur that Grahame's writing
swerves unsteadily between pastoral, farce and didactic children's fare;
the reader is sometimes addressed as if they are an adult, at other
times as if they are a child; and the story is at least three different pieces
of work fairly crudely pasted together. But despite being as uneven as
a badger, it is also a triumph. As a child, I never noticed its inconsis-
tencies, or downright odd chapters; all I knew when I was reading was
that I had completely disappeared into Grahame's Badgerland.

Readers, rather than critics, felt the same, and *The Wind in the
Willows* had been reprinted thirty times when A. A. Milne's theatrical

adaptation, *Toad of Toad Hall,* was first staged in 1929. One of the most enthusiastic early readers was Theodore Roosevelt, the manly American president who gave his name to another cuddly hero, the teddy bear. Writing a personal thank-you letter to Grahame from the White House days before standing down in January 1909, Roosevelt said he regarded Grahame's creations as 'old friends'. As well he might, because his family kept an American badger as a pet in the White House. Roosevelt, who had accepted it as a gift from a small girl during one of his grand tours of the Wild West, memorably described the badger, which he named Josiah, or Josh for short, as 'a small, flat mattress, with a leg under each corner'. When this bottle-fed mattress nipped too many prestigious ankles, it was donated to the Bronx Zoo.

The Wind in the Willows spoke to the wanderer in the American president. 'Indeed, I feel about going to Africa very much as the sea-faring rat did when he almost made the water-rat wish to forsake everything and start wandering,' he wrote to Grahame. Shortly after completing his presidency, just as he had hinted to Grahame, Roosevelt wandered off on a year-long safari that began in Mombasa, took in the Belgian Congo and ended in Khartoum. With an entourage of 250, including porters, tent boys, horse boys and uniformed native soldiers (from several tribes, 'to minimise the danger of combination in the event of a mutiny,' wrote the ex-president), Roosevelt and his son Kermit slaughtered 512 animals, including seventeen lions, eleven elephants and twenty rhinoceros. These gung-ho trophy hunters were accompanied by Sir Alfred Pease, an English MP and badger enthusiast who wrote the first natural history of the

Eurasian badger. Roosevelt described Pease as 'a singularly good rider and one of the best game shots I have ever seen'. Did they sit around the camp fire and discuss the pleasures of hunting the badger in little old England? Later in the expedition, Kermit entered a bamboo thicket and killed a ratel, or honey badger, a fearsome mustelid that is more pine-marten-on-steroids than badger and will launch itself at everything from a pride of lions to the testicles of a rival male. Roosevelt noted it was 'an interesting beast'.

It is difficult today to fully perceive the radical character of Grahame's Badger when his hero has become such a familiar figure in children's literature. John Clare's poem admired the stoicism of the solitary badger pitched against the village rabble but badgers most commonly cropped up in folklore as objects of fear, with something witchy about them, like owls. Grahame's Badger was a completely new interpretation. Given its animal universe, *The Wind in the Willows* encourages us to see Badger not from a human perspective but as other beasts see him. Here, Badger is neither vicious nor persecuted but the rock on which other weaker and less reliable animals could depend; a figure of impeccable integrity and calm authority who is the only animal powerful enough to subdue the rebellious weasels and stoats and bring the arriviste Toad into line. C. S. Lewis put it most dramatically: 'Consider Mr Badger ... that extraordinary amalgam of high rank, coarse manners, gruffness, shyness, and goodness. The child who has once met Mr Badger has ever afterwards, in its bones, a knowledge of humanity and of English social history which it could not get in any other way.'

Literary critics have obsessed over the precise social class of Badger. Is he a lord or a squire? Is he even a gentleman, with his scruffiness and eccentric manners? Badger can treat the working classes brusquely. When a chauffeur delivers a new car for Toad he is hustled away by Badger with the words: 'I'm afraid you won't be wanted today. Mr Toad has changed his mind. He will not require the car. Please understand that this is final. You needn't wait.' Jan Needle's 1990 novel *Wild Wood* cunningly inverts the class politics of *The Wind in the Willows* by retelling it from the perspective of the oppressed weasels and stoats, and we see how hard it is for Baxter, the poor ferret chauffeur given such short shrift by Badger: 'I tried to speak to the badger, tried to argue, like. But it was no good. He turned on his heel and disappeared, slamming the door behind him.'

Badger, judged the critic Margaret Blount, was an unstoppable 'animal version of God and squire mixed into one', while the US critic Lois Kuznets saw Badger as 'an older breed of country squiredom' with his 'potential for antisocial gruffness'. If Toad represented the arriviste endangering social stability by provoking working-class uprising, then Badger was a member of the upstanding landed gentry upon whom fearful folk like Grahame could still depend. The funniest interpretation of class in *The Wind in the Willows*, and how the book shaped our feelings towards real badgers, was offered by Simon Hoggart, the *Guardian*'s parliamentary sketch writer, who was struck a few years ago by the fact that MPs' opinions on a cull of badgers divided so neatly down party lines. Badger, he wrote, was a kindly figure whose purpose was to stop Toad making an idiot of himself. In short, he is 'one of literature's greatest spoilsports' and this is why public school Conservatives – who instinctively identify with childish, rebellious Mr Toad – hated badgers and wanted them dead. Badger, thought Hoggart, would strongly disapprove of the Bullingdon Club, the raucous university drinking society joined by generations of posh boys who went on to become top Tories.

Alan Bennett depicted Badger as a gently homosexual housemaster in his stage play of *The Wind in the Willows*, and no wonder. Grahame's story screams of his fear of women. He spoke privately of creating a fantasy world devoid of the problems of sex, and the first female character does not appear until chapter eight. The only two women in the story, the gaoler's daughter and the washerwoman, are not even given the dignity of a name. Although compassionate towards Mr Toad, the gaoler's daughter is described as a 'wench' who wants to domesticate this wild animal, planning to 'make him eat from my hand, and sit up, and do all sorts of things'.

Critics have identified the inspiration for Badger as everyone from Marcus Aurelius, the Roman emperor and writer whom Grahame discovered at public school, to Grahame's friend Frederick James Furnivall, the co-creator of the *Oxford English Dictionary* and a wild man of letters. In his seminal biography of Grahame, Peter Green thinks that the greater part of Badger is actually Grahame himself: Badger (and Otter) pop up and vanish again without apology, and animal etiquette forbids any criticism of such sudden disappearances; Grahame wished he could be so cavalier in his relations with society.

Humphrey Carpenter argues that Mole, Rat and Toad all, at various points, represent 'the wanderer', a man like Grahame's father, who shrugged off all responsibility for his family and escaped society. Grahame, too, sought to flee the city and his unhappy marriage, and frequently went travelling alone. At the other pole was Grahame the home lover, and this produced Badger's splendid and yet cosy underground quarters. Badger is brave enough to be an adventurer and grounded enough to be a home lover, unlike the unreliable Rat, the timorous Mole and the erratic Toad. Perhaps Badger's rootedness and his ability to slip away from society and yet live within it on entirely his own terms represent not how Grahame was, but the kind of man he wished he could be: a fully formed grown-up. 'Badger is above (or rather, beneath) society,' writes the critic Peter Hunt. 'He is the father figure that Grahame was looking for, the earthy stability that could empathize with the child in freedom of speech.' Carpenter's final interpretation of Badger is even bolder. 'He is the still centre around which the book's various storms may rage, but who is scarcely touched by them,' he argues. 'He is, one may surmise, the deepest level of the imaginative mind, not easily accessible; perhaps he stands for inspiration, only visiting the artist when it chooses, and then behaving just as it wishes.'

The Wind in the Willows closes with mother weasels finding that the ultimate sanction for fractious children is the threat of the 'terrible grey' Badger. 'This was a base libel on Badger, who, though he cared little about society, was rather fond of children,' wrote Grahame in the book's final sentence, 'but it never failed to have its full effect.' In real life, there was no Badger-saviour for either Grahame or his troubled son. The strange cocktail of stories, indulgence and absence that Grahame offered Alastair seemed only to cause contempt and suffering. When Mouse was eleven, his nickname for his father was 'Inferiority'. At nineteen, after repeatedly failing exams at Oxford University, Mouse was found dead on a railway line. His injuries suggested he lay across the rails and waited for a train to decapitate him.

For all the theories of Badger being an ideal father figure, he is also a real wild badger.

Amidst their preoccupation with identifying Badger's precise place in the English class system, even the critics notice 'something Badgerish about Badger' as Peter Hunt puts it. Badger, although rarely visible, seemed 'to make his unseen influence felt by everybody about the place,' wrote Grahame, capturing the traces badgers etch into the countryside. His evocations of Badger's grand home, elusiveness and tenacity are absolutely true to the spirit of the species. Of course there is poetic licence – Badger puts up a couple of lost hedgehogs in his spare room when a real badger would deftly peel back their skins and gobble them up in less time than it takes to unfurl the guest laundry – but not as much as contemporary readers might assume. Badger lives alone, which might seem improbable in Britain today, but Grahame was writing when the badger population was at its nadir; solitary badgers were probably all he ever saw.

Early reviews of *The Wind in the Willows* that criticise Grahame for the inaccuracy of his nature writing seem hilarious today because they miss the allegory. *T.P.'s Weekly* referred to numerous incidents which would 'win no credence from the very best authorities on biology', while *The Times* scoffed: 'As a contribution to natural history, the work is negligible.' And yet Grahame writes beautifully about the countryside. Mole, for instance, in winter, admires 'the country undecorated, hard, stripped of its finery. He had got down to the bare bones of it, and they were fine and strong and simple.' A rare positive review in *Vanity Fair* understood this. 'The book for me is notable for its intimate sympathy with Nature and for its delicate expression of emotions,' judged Richard Middleton. Grahame's portrayal respects the integrity of wild species, as he suggested he would do when he criticised how animals were treated in fables.

An American scholar called Clayton Hamilton met Grahame some years after *The Wind in the Willows* was published and quoted him saying:

As for animals, I wrote about the most familiar in *The Wind in the Willows* because I felt a duty to them as a friend. Every animal, by instinct, lives according to his nature. Thereby he lives wisely, and betters the tradition of mankind. No animal is ever tempted to deny his nature. No animal knows how to tell a lie. Every animal is honest. Every animal is true – and is, therefore, according to his nature, both beautiful and good. I like most of my friends among the animals ... come, and let me show you.

It sounds like something John Clare might have said.

As Grahame's dismay at the rapidly changing human world deep-
ened, so he 'became convinced that he belonged to the world in
general rather than specifically the human race,' suggests Alison
Prince. Humans, Grahame felt, were interlopers in nature. In an
impassioned essay about his need to escape the din of London for
places where his inner ear could perceive small sounds, he wrote:

> This pleasant, many-hued, fresh-smelling world of ours would be
> every whit as godly and fair, were it to be rid at one stroke of us
> awkward aliens, staggering pilgrims through a land whose pleas-
> ant places we embellish and sweeten not at all. We, on the other
> hand, would be bereft indeed, were we to wake up one chill
> morning and find that all these practical cousins of ours had
> packed up and quitted in disgust, tired of trying to assimilate us,
> wearying of our aimlessness, our brutalities, our ignorance of
> real life.

The first fictional badger of any standing to follow Mr Badger was
created by his equally popular contemporary Beatrix Potter. Peter
Rabbit was a huge hit before *The Wind in the Willows* but her badger,
in *The Tale of Mr. Tod*, appeared four years after Grahame's book.
Interestingly, Potter felt Grahame made his animals too human, particularly when Toad combed his hair. 'A frog may wear galoshes; but I
don't hold with toads having beards or wigs! So I prefer Badger,' she
once wrote.

Her villainous badger, Tommy Brock, was more in keeping with the prevailing view of the species at the time than Grahame's. Brock, 'a short bristly fat waddling person with a grin; he grinned all over his face', was 'not nice in his habits. He ate wasp nests, frogs and worms, and waddled about by moonlight, digging things up. His clothes were very dirty; and as he slept in the day-time, he always went to bed in his boots.'

Potter spent much of her life in the Lake District, close to badgery places such as Brockhole but in a region from which badgers were almost completely exterminated at the time she was writing. 'No one else grubs up the moss so wantonly as Tommy Brock,' she wrote. Was she influenced by the prevailing views of local farmers? Were her lettuces pinched by badgers? Potter conceded that Tommy Brock only occasionally ate rabbit pie but she still made him imprison a whole generation of bunnies in his oven. In her clever, sly portrait, Brock is a common labouring man, unlike the gentlemanly fox, and carries a spade, rather like a real-life badger digger. It is not the fox who is cunning, but duplicitous Brock: invited into a rabbit burrow by old Mr Bouncer, the badger smokes one of the rabbit's cabbage-leaf cigars and then makes off with a bundle of seven rabbit babies. He lies to Peter Rabbit that his sack contains caterpillars and proves craftier than Mr Tod when he feigns sleep as the tricksy fox tries to rig up a pail of water over his bed.

Aspects of Tommy Brock – sleeping in the day, grubbing around, eating rabbits – are accurate enough. Like all wild animals, the badger is neither good nor evil; it is we who imbue its struggle to survive with a moral character – bad Tommy Brock or good Mr Badger. And yet Potter's negative vision of the badger gained almost no traction at all in the subsequent decades of storytelling, whereas Grahame's version did.

Grahame's Badger inspired many cetes of other inspirational fictional badgers, from the heroic Trufflehunter in C. S. Lewis's *Prince Caspian*, published in 1951 as the second of his Chronicles of Narnia, to Bill Badger, Rupert Bear's best friend, in the comic strip that began in the *Daily Express* in 1920. Most are pale imitations of Grahame's subtle Badger: occasionally curmudgeonly male authority figures with a nod towards their habits in the wild; 'human beings with animal heads', in the words of the critic Tess Cosslett. In the 1950s, Tufty Fluffytail the squirrel was created for the Royal Society for the Prevention of Accidents to publicise road safety for children, alongside Policeman Badger, who always saved the animals in the nick of time. The badger in Colin Dann's 1979 novel *The Animals of Farthing Wood* is another dependable leader of the pack. The sadness of animal society when a badger is dying, in Susan Varley's *Badger's Parting Gifts*, has become a popular reading at funerals. Varley said she chose a badger because 'it's a strong, sturdy-looking animal', but admitted its character was inspired more by her grandmother than a real animal. Detective Inspector LeBrock, a magnificent beast who wraps his muscular torso in a Sherlock Holmes-style coat in Bryan Talbot's splendid *Grandville* steampunk graphic novels, published since 2009, is slightly less a pillar of the community. In Talbot's world, populated by foxes, boars, turtles and rats, human beings are 'dough faces', a forgettable 'hairless breed of chimpanzee' who are slaves to the powerful animal. LeBrock encounters various thugs and a totalitarian regime but still emerges as the violent, brave hero.

Was it a coincidence that the badger's revival in the British countryside began in the years after the publication of *The Wind in the Willows*? Attitudes towards the animal were undoubtedly beginning to change before Grahame's story. Angela Cassidy found the first positive mention of a badger in *The Times* in 1877, when a letter to the editor extolled the fact that 'no creature is more cleanly in its habits'. The early decades of the twentieth century witnessed a great flowering of nature writing: the prose of Henry Williamson; the poetry of Edward Thomas, who described the badger as 'that most ancient Briton of English beasts'; and countless less acclaimed works of natural history. Sir Alfred Pease described his account of badgers, published on the eve of the First World War, as 'a plea for tolerance, if not affection for the species which has been so grievously wronged at the hands of man for centuries'.

These new naturalists were not effete badger cuddlers – they often applauded the species for utilitarian reasons. 'It is always a pleasure to record the occurrence of the much-persecuted badger,' wrote the *Manchester Guardian*'s country diarist in 1906, lamenting its absence from much of the country. In the first decades of the last century, the paper's country correspondents regularly noted their helpful ability to destroy wasps' nests and remarked on what good pets they made. Occasionally, a defence of the badger was put in most un*Guardian*-like terms. In 1923 a diarist called T.A.C. approvingly quoted the explorer Sir Harry Johnston, who opined that badgers were slaughtered 'for no earthly reason but the unreasoning love of destruction which classes us as a race on a level with the worst type of negro'.

Attitudes towards the badger were moving in Grahame's direction but broader social transformations also helped save the badger. Years of agricultural depression caused by cheap grain imports during the latter quarter of the nineteenth century reduced the land under cultivation by a third, lessening the pressure on places where badgers lived. The First World War wrenched young men from the countryside and the social and economic upheavals that followed saw the decline and breakup of many great estates. This dramatically reduced the number of gamekeepers, so control over 'vermin' like badgers slackened.

The modern conservation movement with its nature reserves and its new science of ecology began in the second decade of the twentieth century, when an increasingly suburban society began to worry about its losses. A growing tendency to view wild animals with wonder meant that badgers were now watched and studied rather than baited and trapped. The transformation in the fortunes of the badger might have happened without Mr Badger, but I bet it would have been far less pronounced. With Kenneth Grahame's help, the badger became lodged in the popular imagination as a dependable, morally upstanding animal, part of a trusted iconography of Britain. In heraldry, the badger is both the star of the House of Hufflepuff's coat of arms in *Harry Potter* and decorates the coat of arms of Tesco, which has its headquarters close to Broxbourne (which means 'badger's brook') in

Hertfordshire. The badger is also a symbol of conservation as the chosen image for The Wildlife Trusts charity; a symbol of comfort as a popular brand of real ale; and a humorous staple in viral videos and comedy. (Marcus Brigstocke's Badgerland – 'there's hundreds of badgers all under one roof, it's called Badgerland, Badgerland' – is an execrable theme park.)

The badger is both ancient and modern. Ever since Mr Badger, badgers have been celebrated for being British in character, a grumpy but principled, tenacious, castle-dwelling race who are slow to take offence but mightily hard to shift. As I was to discover, they are also mightily hard to see.

5

Brock

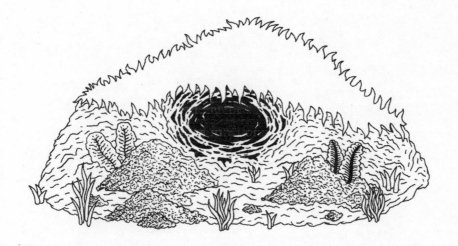

In June 1953, Eileen Soper sat alone in the dark in the middle of nowhere waiting for badgers while everyone else was celebrating the Queen's coronation. 'Fireworks displays started immediately in the villages north and south, about a mile and a half away,' wrote Soper, whose day job was illustrating Enid Blyton's adventure stories. 'I thought this would mean the end of watching for the night, but to my surprise only one cub returned to the sett. He was in no hurry, and was soon out again. I could hear the family wandering about at the bottom of the dell, unconcerned, while Roman candles, rockets, and other spluttering devices shrieked their way up into the sky frightening every domestic animal within miles.'

For Eileen Soper, badger watching seemed to require going against the grain of normal human behaviour. In fact, the more I read about badger watching, the more it seemed that its exponents were often perched awkwardly outside society, looking in, or wishing the ordinary world would quieten down and stop frightening the badgers. It felt like I too was required to cast aside convention to find badgers. Part of me delighted in abandoning a humdrum existence but part of me felt fearful. Several years before I had dedicated a year to an unfulfilled passion from childhood and sought to find every species of British butterfly. My devotion to that cause cost me my relationship with my girlfriend, Lisa. Now we were back together and I was embarking on something even more antisocial: a nocturnal, monkish pursuit that was difficult to share with others and demanded many nights away from home.

After reacquainting myself with the countryside at night but failing to find any badgers at Wookey Hole, my next attempt coincidentally echoed Soper's coronation badger watch. I had been reporting on street parties in the West Midlands held to celebrate the marriage of Prince William and Kate Middleton, and when I checked into a pub in Bewdley, the small Worcestershire town was strewn with the entrails of a party that had lasted all day. Three lads wearing matching T-shirts saying 'Wills + Kate 4eva' crossed the road. Children with painted faces ran along the pavement. I changed into dark clothes and left this party, driving out of town on empty roads. Heading into Shropshire, I parked at a farm where a series of long slender fields rolled gently down to the wooded banks of a tributary of the Severn. I had permission from the farmer to be there but still felt a pang of illicit pleasure, and wondered how I would explain myself to anyone else who was out walking rather than toasting the new Duke and Duchess of Cambridge. The innocence of badger watching had been tarnished in 2003 by the Labour politician Ron Davies, who, when confronted skulking in the shrubbery at Tog Hill, near Bath, by a *Sun* journalist, had uttered the infamous excuse that he was not looking to perform, in breathless tabloid parlance, a lewd sex act but was searching for badgers. Davies had unwittingly chosen an apt story, for 'badger game' is actually a piece of American slang describing the sexual entrapment of respectable men in order to blackmail them. Unfortunately for him, it was also broad daylight and the newspaperman thought to check with local badger experts, who confirmed there were no setts in the area. For a while it appeared that 'badger botherer' might challenge 'dogging' in the lexicon of contemporary behaviour in the countryside.

It was a glorious April evening when I set out and the air was choked with birdsong, the aural equivalent of a sky full of stars, or an overcrowded FM frequency. I tuned in and out of competing blackbirds, blackcaps, robins, thrushes and the raucous kark of a pheasant. A curlew overhead sounded a melancholy note that spoke of coastal wetlands rather than rural Shropshire.

The sweet smell of muck spread on a distant field rose from beyond the darkening copse, and the breeze rattled the boughs of a rotund oak. An ancient hedge, rich in blackthorn, hazel, lime, elm, wild rose and elder, led me to the valley bottom. Creatures whizzed away, no matter how softly I trod; pigeons crashed out of ash trees; rabbits stood alert, as motionless as garden ornaments, before speeding off; I even woke a roosting pheasant. Warning calls rang out, a conspiracy against me.

There were no public footpaths across this valley but there was a wavering line through a green field of newly sown barley. I had now acquired rough theoretical knowledge of badgers and knew this path had been made by the creatures. Searching along the field edge, I found five badger latrines, neatly dug and filled to the brim with dark pyramids of poo. Where the slope closed in on the stream lay a rug of bluebells divided by another wobbly badgerway running towards a sett. A rabbit hole is round, like a rabbit. Badger holes are more crescent shaped: a flat-bottomed tunnel, broad and low-slung as if a badger has cast her body in soil. Two sandy spoil heaps from the tunnelling were as bright as coral cays in this blue sea. Badgers and bluebells were thought by some to be symbiotic. The writer Roger Deakin recalled Suffolk friends who owned Tiger Wood – so called because the canine of a sabre-toothed tiger had been found there – who believed their plentiful bluebells were thanks to the many badgers who beat down the bracken.

I waited in the shadow of a bent old ash, the only tree yet to show its leaves. Sitting by a limb of broken branch, still warm to the touch, would, I hoped, help conceal me. A duck, a rook, a dog, a car. Noises became more distinct as my surroundings became more indistinct. A stoat scampered across the track with a bouncy, alert run, and colour seeped from the scene. Ashley Ford, a Norfolk writer, noticed that night does not fall but rises. It begins with the earth: grass stems, roots, soil, stones, twigs, gravel, all lose their individuality, blur and become shadows. The dusk rises to the trees, and the shadows cast by barns and buildings. The sky is the last to go, still bright blue at sunset, then navy, then grey, and finally black.

By now, the bluebells had disappeared in the darkness but the may blossom continued to shine bright white and the two sandy spoil heaps below the badger holes showed up like lights. If anything emerged, it would be picked out on these spotlit patches.

I could hear gusts of wind moving across the field towards me before they arrived. Periodically they fetched up the sweetness of pollen from the bluebells. When I pulled up my hood to conceal my face – badger watchers had warned me that the face becomes a conspicuous moon at dusk and exposed hands waggling about are as visible to badgers as a mime artist's white gloves – I realised how quickly I had become dependent on my hearing for a sense of what was unfolding around me.

Then, a scream. A baby rabbit? An owl? Or a meeting of the two? I could not tell if the cry was predator or prey. I realised that although I could claim a basic familiarity with the natural world by day, the night was alien to me. From the woodland below came a noise which I felt sure must be the whickering of a badger. It was a puppyish sound, half-dog, half-fox. There was a rustle and some squeaky barks but, frustratingly, the source of the sound remained within the trees, out of sight.

It was dark and late and I concluded that I had blundered around far too much to see a badger. There was, at least, something deeply soothing about being alone in the countryside at night. My eyelids gained weight. As I floated between wakefulness and sleep, I savoured the novelty of lying on a hillside in the dark when Prince William and Kate would be cutting their cake and clasping hands for a first dance, and millions raised their glasses in warm pubs across the land. A few hours later, drugged by my evening alone outdoors, I climbed into my car and drove uncharacteristically slowly through Wyre Forest, expecting to see a badger on every corner. I did not meet a living thing until I entered Bewdley again and encountered the wobbly celebrants of the royal wedding, weaving their way home after a long day in the pub.

My next search for badgers was serendipitous. My heart had sunk when I was allotted accommodation for the Hay Festival: on the map, my lodgings looked a long way from the little town of books on the Welsh border. When I arrived at a rather grand Victorian Gothic house of grey slate and stone just off the road to Builth Wells, however, it was all rather lovely. Rhododendrons were plumped up like pillows in the gardens, the air smelt of woodsmoke and there were wellies scattered in the porch like pins in a bowling alley. Best of all, my kind hosts revealed their wood was rife with badgers, and they seemed to regard my decision to spend Saturday night in search of them as the most natural thing in the world.

I was coming to understand that finding a badger was not as simple as standing in silence downwind of a sett. It required meticulous preparation. People deployed all kinds of ruses to confound the olfactory abilities of the badger. One experienced watcher rubbed her hands in soil and on smelly elder leaves; another I met left his coat by a badger sett for several weeks, then collected it when it was adequately marinated in the mossy tang of the outdoors. This stratagem didn't work with his gloves because the badgers snaffled those for bedding.

Finally, a few dedicated souls won over cautious badgers simply through persistence, turning up every night for a year, or more, so that eventually the badgers were not merely unconcerned but would actually scent-mark their boots and trousers, a sign of a person's acceptance into the animal's social group.

As I set out across the paddock beyond the garden, I realised I had made a basic mistake. An hour before, in Builth Wells, I had devoured fish and chips, and my hands, to a badger's sensitive nose, would reek of fishy vinegar. I bent down and scrubbed them – in earth and then with the green-staining soap of fresh fronds of bracken. I was still not sure if unwashed was as bad as washed, although at least I had learned not to apply aftershave before a night seeking badgers.

Now, twilight was tiptoeing in. A large grey stone materialised in the centre of the field and only when I got closer did I see it was a dead sheep. Entering the wood over a gate, I felt I was being watched. There, on the rough track ahead, stood a tawny owl, looking like those mottled stone ornaments that clutter the less visited corners of garden centres. It gazed at me with the contempt of a cat and then, after a full thirty seconds, beat its wings silently away through the trees. Odd, that some long-dormant natural instinct in me had correctly sensed the owl's presence.

Gerald of Wales, the medieval chronicler of border country, imagined a world of badger kings with armies of subterranean badger slaves when he rode through these valleys 800 years ago, and they were well populated by badgers once again. Within the gloom of the wood, obvious badgerways zigzagged through the undergrowth. A concern to avoid cracking twigs made me walk as if I was on the moon. I felt I was becoming part of the night when I spotted a raven gurgling happily to itself, completely unaware of me. Only when I arrived directly underneath its perch did it see me, call out in surprise, and fly off.

I climbed a tree that looked over a convergence of badger paths resembling Spaghetti Junction, carved into a thin carpet of dog's mercury. An elevated position is a good spot to watch any animal; just ask a bird of prey. The only problem is the absence of a comfy old tree by every sett: that fantasy oak into which you can clamber with ease, swing your legs and gaze down on gambolling badgers. When you are a child every tree looks climbable; is it a failing of the joints or of optimism that makes it so difficult to identify suitable climbing trees as an adult? After a brief, unseemly struggle, I established an uncertain position at an altitude of eight feet. Being propped between two branches was comfortable for about ten minutes. As pheasants and pigeons began their pre-roosting rituals, I discovered how badly adapted we are to spending an evening on a branch.

The paths stayed empty of badger traffic and, as it grew darker, I gave up on my torturous roost. As I tiptoed through the wood, fervently trying not to crack a twig, I disturbed something, and it took off down the slope. A grey-black shadow, a humped back, shifting at surprising speed. It had to be a badger. It *was* a badger. Definitely. Far more dog-like than a deer, it bounded away and was gone.

Follow a badger motorway in the right direction and, just like our main roads, it will widen as it nears the metropolis. I found a squiggly track that must have been established by badgers because it dived under obstacles that a taller mammal, a deer or a man, could not easily pass. After a while, it broadened into a vast dirt living-space, an adventure playground beneath an overgrown coppice. This was not for forest imps – that would imply these creatures were slender – but for woodland dwarves: podgy, padded, frolicsome beasts. There were old tree stumps for jumping off, dangling branches for scratching, and a steep slope to slide down. So well trodden was its floor that nothing grew here. In the very centre were three entrances to a sett, sandy earth cascading from them, patted smooth by the passage of many paws.

The wood was wet. A dull grey evening seeped through the bracken, bramble, budding honeysuckle and last month's bluebells. Their pods of green seeds stood tall but the damp left their floppy leaves as slippery as ice. The loss of twilight was so gradual that it seemed as if it would never get dark. Sounds thinned out. The world was winding down. Lambs complained in the field. After a long silence broken only by the occasional swish of a car on the road to Builth, the rooks started off again, chuntering for ten minutes in a desultory fashion, the last conversation before lights out.

Finally, silence. There was no movement around the setts. No nose ascending. Could the badgers smell the ink on my pen as I scribbled notes? At night, sitting still, alone, we become far more acutely aware of our own human sensory footprint. When I yawned, I could perceive the warm stale air I emitted, weirdly pungent and fleshy beside the understated scents of soil and moss.

With a sigh, the rain arrived. Its droplets rarely reached the ground beneath the canopy as outstretched leaves intercepted them with a flick and a tap. After more than an hour I was so comfortable that, once again, I could feel myself dozing off. Then, finally, an animal. High up the slope, above the badger setts; small, furry and black against the horizon. Suddenly alert, I watched it as intently as I could out of the corner of my eye, as recommended by the writer-poacher Richard Jefferies. The animal jigged about for a bit, impatient for playmates, and then bounded away with a cat-like leap. Not a badger but a fox cub. Foxes were lodging in the premises.

Dead branches became fingers; a frond of ivy turned into a snake; and I skidded into a dream in which a badger walked past, two metres away, looked at me evenly and decided I was not worth fearing. I woke when I felt cooler air sinking past my face. The leaves on the trees above were blots of black ink running into the dark-grey sky. Everything was growing smudgy. After two hours half watching, half dozing, I had failed to see a badger emerge. Foxes and badgers may sometimes live together in a large sett but the presence of fox cubs here made it unlikely I would see a badger.

I got up, with a crack that sounded like my back but was a rotten branch beneath me, and moved off like a sleepwalker down the steep hill. When I emerged from the wood, again passing the silhouette of the dead sheep, it felt so light that it could have been dawn. The grand old house was a grey bulk against the darker trees, its windows blank except for one small lighted square: my room. Before I fumbled the latch, I stumbled across a final nocturnal treat. In the grass shone a brilliant green LED light. Then I saw it was a glow-worm. When I picked it up, its body curled around the light but it was never fully extinguished. The only thing I knew of these insects was what I had read in *James and the Giant Peach*. She was a female, and her glowing beacon of light was an invitation to any males that flew past.

Since I had had little luck locating badgers in the evening, a few weeks later I decided to try the morning. My father, John, lived in Devon, a land extremely well populated by badgers, and during a brief stopover at his house in Ashburton I set out in the dark, at 4 a.m. As I walked up the steep lane out of the town, the sky lightened to reveal a vaulted ceiling, a great expanse of high cloud fractured like old plaster. Imperceptibly, it was breaking up, like some slow-moving kaleido-scope, illuminated by the moon in the western sky and the promise of the sun from a slim band of yellow on the eastern horizon.

The world was silent, except for the episodic rumble of HGVs, lit up like Christmas trees, on the A38 in the valley bottom. It appeared that dawn in July would have no chorus. The birds' territorial battles had been won, or lost, and now they were hunkered down to rear their young. The lane narrowed and at the point where the houses and tarmac ended, it dived between thick Devonian hedge banks and became a little-used path, an overground tunnel of hazel and black-thorn. All I could smell was vegetation – a dewy, earthy green mulch.

The first living thing I saw was a bat, darting under the vaulted ceil-ing, and I paused as it echo-located around me, checking me out, sonically at least. The band of pale yellow became orange and grew broader and I was surprised to see the pen I was writing in was blue, not black. Colour was seeping back into the world. I had become attuned to the pattern of the dusk and now events seemed to be unfolding the wrong way round; the world getting lighter felt like time moving backwards.

I was heading for the Field. My dad and other neighbours had recently clubbed together and bought a hilltop paddock on which they were planting trees, grazing sheep and growing vegetables. There were two caravans, a tepee, and that lovely tidy untidiness of allotments: runner beans on canes, furrows of potatoes and stupendously fat onions. It was a bountiful year, the first season for decades in which anything had been required of the Field's well-rested soil.

The badger sett at the bottom edge of the Field was invisible; here, the sheep had been defeated and the meadow blurred into a rising tide of bracken and bramble and stringy nettles, as tall as me and bitter with stings. I sat downwind of the best-looking badger path and waited.

One bird started up, a thrush; the engine of a farmer's Land Rover chuckled in a nearby lane and the church bell in Ashburton tolled five. The belly of the remaining cloud was now tinted with pink. A blackbird landed on a stag-headed oak that stood above the sett. The oak was suffering, its limbs dying and dropping off. Had badgers dug away its roots? At 5.10 a.m. the first swift of the day sounded, the blackbird bobbed its tail and shrieked chup-chup-chup-chup and a wood pigeon cleared its throat for a woozy woo-woo, with that careless, unfinished note at the end.

Then, in an explosion of movement, something crashed through the undergrowth in front of me. The noise disappeared, as suddenly as an insect might be snatched out of the air by a lizard, right by where a badger might enter one of the hidden holes. The sound was more like a bolting deer than a clumsy, foraging gait, but if it had been a deer, I would have seen it through the nettles. Badger? If so, it had taken sudden fright, having sensed the presence of its ancient enemy: me.

As a local cow began tearing at the hazel that bordered the north side of the Field, the sun appeared, a cupcake of bright pink. It came into view over the eastern horizon so quickly I was struck by a melancholy sense of time moving more speedily than I could ever comprehend. For a moment, life seemed very short. Within seconds of rising, its colour shimmied from pink to orange to a blinding yellow, and then dawn was done.

I headed for home, past a huddle of sheep gathered in a loose circle like exhausted revellers at a festival. Paranoid, the sheep turned their necks and followed me with their gaze as I loped past. Around the next corner, a roe deer bounded away, having detected me long before I saw it. My enfeebled senses when placed in close proximity to prey made me feel a very poor excuse for a predator. When I got back to Dad's house, I fell into a jetlagged sleep until the church bell chimed the half-hour past eight.

6

Bodger

I can still recall the thrilling acrid pong of Grandma's garage. When Grandma and Grandpa were otherwise engaged, my sister and I would tiptoe across the concrete floor in the gloom, raise the great lid of the chest freezer, and stare, repelled and fascinated, at the yellow chicks, sticky with frost, that lay inside. Twice a day, the dead chicks would be extracted from their chilly coffin, thawed and fed to Rusty and Dusty or Wildone and Wildtwo. These were injured tawny owls, which Grandma devoted her days and nights to nursing back to health in home-made aviaries in her garden, alongside foxes, hedgehogs, barn owls, buzzards, little owls, short-eared owls, sparrowhawks and kestrels. The kings and queens of this ever-changing menagerie, though, were the badgers.

Jane Ratcliffe's labours were not a business, nor a charity, and she was not a vet. She was a home economics teacher and stay-at-home mum who often chose the company of animals over humans in her life. She had arrived at her mission to save injured creatures and return them to the wild by chance, when she watched badgers in the countryside around her home in Cheshire and witnessed the havoc wrought by diggers and baiters. She began looking after orphaned and injured badgers, and joined a growing campaign to force Parliament to pass a groundbreaking new law, making the badger the first wild land mammal in Britain to be given protection from persecution. Kenneth Grahame may have been responsible for the dominant image of the badger in the twentieth century but it was a small group of unconventional women like my grandma who changed our relationship with badgers for good.

The first of these women to depict badgers in a radically different light was Frances Pitt, whose charming memoir *Diana, My Badger* was published in 1929. The well-to-do Miss Pitt recounted how one day the local rabbit catcher delivered two baby badgers to her, which she called Diana and Jemima, and reared by hand. Pitt was an old-fashioned kind of animal lover who also loved to hunt foxes but she decried the plight of the badger – which 'owes nothing to sport, like the fox or the deer, and has always had to fend for itself' – and showed great affection for her adopted charges. 'Diana's attitude was that of a well-brought-up young dame, and she wore a "I don't speak to strangers" air that was discouraging,' wrote Pitt. Diana befriended the family dogs, Bogie, Jack and Nettle, despite the fact that the latter was the type of terrier bred to dig badgers from their setts. She expressed her feelings through her wiry grey coat. 'Ripples of excitement ran through her, her fur rising and falling with the emotion, special excitement being shown by a fluffed out tail,' Pitt went on. 'Her ridiculous little short white tail would then spread out like a bottle brush.' When Pitt's pets ran away from their human home to a nearby wood, she became a badger watcher instead, and enjoyed seeing Diana bring up three cubs, in the wild, as she should.

Pitt's book probably inspired thousands of people to take badger cubs as pets. Cubs were often discovered alive but helpless after badger digs and were routinely sold through classified adverts and in pubs until the 1970s, a misguided kindness that could only be cruel for a wild animal. But the magic of Pitt's descriptions also mobilised a generation of nature lovers, including Ernest Neal.

In the decades after the Second World War, Ernest Neal became the most famous badger expert in the land. A biology master at a public school in the Cotswolds, he conscripted his students into helping him observe badgers in a nearby wood and brilliantly wrote up their collective findings in a book called *The Badger*, published in 1948. Neal recounted in his autobiography that he was 'amazed' by the success of his book, which sold 11,000 copies in its first print run alone, helped Neal bring the badger onto television screens in the first BBC wildlife documentaries, and triggered a mini-publishing boom of other amateur naturalists' enthusiastic badger tales, many of which – following in Pitt's footsteps – were written by women. Instead of men drily recounting facts about badger biology or extolling the pleasures of catching, fighting and killing the fearsome creatures, these authors showed readers how it felt to watch, sketch, photograph and save them. Suddenly, people came alive to the possibilities of a non-combative relationship with badgers.

The most compelling tales of how humans formed relationships with wild badgers in the 1950s and 60s were written by two very different but equally unconventional women: Eileen Soper and Norah Burke. Born in 1905, Soper was a child prodigy who at the age of just fifteen exhibited at the Royal Academy in London, where she caught the attention of Queen Mary, who bought two of her etchings. Soper's father, George, was an accomplished artist who was obsessed with germs, and after his death in 1942 Eileen and her older sister Eva, a talented potter, shut themselves away from the world at their home in Hertfordshire, increasingly terrified of disease. Although they fretted that sneezes from strangers could cause cancer, they welcomed wildlife, and children, into their lives. Soper would stop children in the street and sketch them at play, during a twenty-year career producing hundreds of defining illustrations for Enid Blyton's children's adventures.

Over the years, the Soper sisters' house grew increasingly ramshackle and their four acres of garden ran wild. Soper shared her studio with blue tits and great tits, and despite her fear of disease would have these birds take nuts from her lips. 'Those who would share a house with birds must abandon all hope of keeping up appearances. We have scarcely a chair that has not had its upholstery tried out as nesting material at some time,' she wrote. Soper also found solace – and pain – in long nights watching a sett in a half-acre dell close to her home. She filled notepads with vivid pencil sketches of badgers alive with joy and movement and, in 1955, published her book *When Badgers Wake*, which she dedicated 'to the badgers that were kind enough to ignore, sometimes, my association with the human race'.

Eileen Soper writes as if she would prefer to belong to the badgers. When a summer fair was held at the village a few fields beyond 'her' sett, her badger watching was disturbed by 'a jazz band with a crude nasal accompaniment amplified through the loudest and most offensive machine I have ever heard'. In contrast, she revelled in the conversational 'rumbling growls' she heard among the badgers. On one occasion, she found herself apparently welcomed into Badgerland, surrounded by the animals as she stood in the shadow of an elm tree. 'Perhaps the summer midnight had worked its charm on me, or it may be that badger-watchers, in company with bird-watchers, are not entirely human,' she wrote. 'I felt I had encountered a moment of rare good fortune, and that here, in company with those merry wanderers, I had met the fabulous spirit of midsummer night.'

There was no happy ending for Soper, however, for her sett was repeatedly gassed by a local farmer, an illegal but widespread way of killing an animal considered a pest for ploughing up pasture in its pursuit of worms, or taking pheasant chicks reared for a shoot. Eileen's badgers were not immediately killed, and on 31 May 1955 she admired the cubs at play for what would be the final time. *When Badgers Wake* ends with the terse line: 'No badgers survived a final gassing on June 1.'

Born two years after Eileen Soper, Norah Burke, an adventurer and author, saw badgers close to her home in Suffolk meet a similar fate in the 1960s. Raised in the forests of the Himalayan foothills, where her father was a colonial officer in the era when the naturalist and author Jim Corbett was shooting man-eating tigers, her first novel was published when she was twenty-seven; up until her death in 1976, she wrote numerous historical novels, romances and travellers' tales, as well as accounts of her badger watching. She became, in the words of a fellow badger watcher, 'a tough, efficient field naturalist' and inspired awe for her woodcraft. Festooned with undergrowth, she would disguise her scent with bruised elder leaves and pine branches, hold a sprig of leaves in her mouth to break up her face, which otherwise stood out sharply in a wood at dusk, and thereby got within a few feet of the shyest creatures. 'My object,' she wrote, 'is to see the life of these wild animals as it goes on when no human being is present.' Despite her purist stance – like many watchers, she refused to feed wild animals – she got so close to badgers that she once heard one hiccup in the sett.

I warmed to Burke when I read *King Todd*, her tale of a badger she called 'a wild king who, for many years, must have known nothing but starlit peace in his forest, until the day came when he was attacked by mankind'. She admitted to 'boredom as well as irresistible excitement' when alone looking for Todd in pinewoods near Bury St Edmunds, but her writing does not convey any tedium: she describes the wind sailing 'over the trees like surf' and ladybirds crawling across twigs 'like scarlet tortoises', and it all sounds glorious. Burke believed our persecution of the badger was an ancient impulse. 'There has come down to us in our blood from prehistoric twilights where every shadow might contain death,' she wrote, 'a fear of all wild creatures, and enmity towards them.'

Unlike Soper, Norah Burke did not retreat into her Badgerland but was an engaged and prescient critic of the way we managed the countryside, predicting the rise of the EU stewardship schemes that pay farmers to manage their land for conservation, the reverse of the incentives being offered to destroy hedges and industrialise farming in the second half of the twentieth century. 'If we want full bellies, we must accept modern methods of agriculture; and if the Nation wishes to preserve England's green and pleasant land, we must pay farmers to keep their hedges, instead of blaming and squeezing them from every side,' she wrote in 1963. 'The money must come from your pocket and mine. It's as simple as that.' Perhaps more so than ever in the era of the big supermarkets and cheap food, we are still wrestling with this uncomfortable truth.

By the 1960s, as Burke noted, some of those who managed the countryside recognised the merits of the badger. Foresters began to incorporate badger gates into rabbit-proof fencing around conifer plantations. These sturdy ten by fourteen-inch doors were like a cat-flap: hinged at the top, they swung backwards and forwards. Only the badger was bullish enough to ram its head against the flap and push its way through, and the gates stopped them tearing holes in the fencing and enabled them to move freely through the plantations, devouring pests such as slugs, snails and rodents. The badger gate was a pleasing example of an accommodation between humans and badgers that benefited both species. It also became a symbol, for my grandma, of how wild animals must be given their freedom.

*

The daughter of a dentist, my grandmother Jane Ratcliffe grew up in industrial Yorkshire with her twin sister, Joan. She fell in love with wild things one summer holiday when, as a small child, she set out, alone, to go fishing in a nearby pond. To her surprise, she caught two sticklebacks. Marvelling at their fiery red breasts, she kept them in an aquarium and watched them make nests. From then on, nature offered her an escape, as well as a sense of purpose, expertise and authority she was lacking at home. Jane's mother was glamorous and not particularly maternal; her father was a typical Edwardian parent, remote, strict and yet also periodically indulgent of his twin girls.

Joan was taller and always more gregarious, and Jane did not derive much confidence from her family life. Both girls, aged seven, were dispatched to a boarding school in the Cumbrian countryside. At school, Jane organised a country walk with friends every Saturday and when she was ten nursed injured sparrows and a house mouse. One Sunday after church, she found a swift floundering in a puddle and placed it in a cardboard box, which she put on a stool by the oven to warm up. Realising the bedraggled bird would need its plumage straightened before it could fly again, she found a soft-bristled baby's hairbrush and carefully brushed its back and breast, a small act that was characteristic of her instinctive assurance and inventiveness when it came to caring for wild animals. By late afternoon, the rain had stopped and she stepped outside with the swift and threw it into the air. Recuperated, the bird flew off, joining other swifts making darting, screaming passes over the garden.

Jane trained as a home economics teacher and fell in love with Teddy, a clever engineer and the doted-upon son of a publican. They married, and in 1942 their first daughter, my mother, Suzanne, was born. Another daughter and a son followed and, after a war spent flying Lancaster bombers, Teddy became an engineer for Liverpool city council. Jane must have been busy with her young family but she was well organised, full of energy and not particularly maternal, at least not with human beings. 'She always felt very strongly about weak little things,' remembered my mum, 'but I think babies grew up whereas animals didn't.'

By the 1960s, the children had left home and Jane began observing wild animals in earnest. Dividing time between her and Teddy's home on the Wirral and the house they built for their retirement overlooking Lake Windermere in the Lake District, she watched badgers and birds and acquired a reputation locally for nursing wild animals back to health. She always hated cruelty to animals, and her children remember her reporting mistreated horses to the RSPCA. What really mobilised her, however, was badger digging, which she considered 'sadistic, unpleasant and barbaric'. Jane and Teddy volunteered for their local Wildlife Trust and recorded badgers at twenty-seven setts in Cheshire. They enjoyed photographing them foraging at night but were also on the lookout for the terrier breeders who travelled from Stoke-on-Trent and other towns to test their dogs on badgers. Between 1969 and 1971, fifteen of the setts they watched were destroyed, most by badger diggers. By 1973, just one of the twenty-seven was still occupied. Suzanne had vivid memories of Jane's anger: 'I can still hear mum saying, "They come out of the Potteries with their terriers and it's their sport!"'

For all the sentimental feelings inspired by *The Wind in the Willows*, badgers were still suffering in the 1960s. The motorways unfurling across the country cut through their ancestral pathways. Many were killed stubbornly following their traditional routes. Poison continued to be illicitly pumped into badger setts by farmers. Digging had never gone away but had evolved into an urban sport rather than a village activity. Keeping terriers and whippets or lurchers had become particularly popular among the miners of South Wales, Staffordshire, Nottinghamshire, Yorkshire and the north; dogs were a source of pride and status and, like the Victorian breeders, owners wanted to test their 'working' dogs against an animal seen as the ultimate quarry – the badger. Miners were not averse to several hours' hard spade-work on their weekends, either. The diggers made contact with each other through newspapers, which carried adverts for working terriers. 'Lakeland type, 4.5 years, over 150 foxes, countless badgers ... Holmfirth, Huddersfield,' read one. 'Three sensible experienced men require fox and badger work for their terriers. North and North Midlands preferred,' said another.

There was, however, growing popular opposition to this slaughter of the badger. By the 1960s, naturalists no longer sought out badgers in glorious isolation but began forming local groups to stop persecution and go watching together. The New Forest Badger Group was a typical example, formed in 1965 to stop the local hunt blocking badger holes in its pursuit of the fox. Alongside the gathering momentum of a broader conservation movement that was creating county Wildlife Trusts, these new groups began to argue that the badger needed special legal protection.

Parliament and government were adrift of this mood. In the 1960s, an MP even introduced a bill to legalise the gassing of badgers, which was banned under the 1911 Protection of Animals Act. This anti-badger bill was defeated but government mandarins maintained that the animal was not an endangered species and therefore not in need of protection. Ernest Neal recorded in his autobiography how his efforts to raise awareness of the plight of the badger were supported by my grandma and another campaigner called Ruth Murray – both articulate, rather forceful women, who took to news and current affairs programmes on television to campaign against digging and baiting.

Through a friend, Jane met Frances Pitt. She was clearly inspired by her encounter with the author, who lived in a grand house in Shropshire, was master of a fox hunt, and also cared for an ever-changing menagerie of wild animals, including several otters and a tawny owl called Hooter. Some years later Jane named an owl in her care Hooter and declared it was Pitt who had encouraged her to talk and then write about her experiences in two books, *Through the Badger Gate*, published in 1974, and *Fly High, Run Free*, which came out five years later. Jane also became a prolific lecturer, giving talks to wildlife groups and Women's Institutes around the country, showing slides and even playing sound recordings of badgers assembled by Teddy, a practical man who seemed content to play a supporting role to his wife's passion.

As Jane became well known for her work in the North-West, she began to receive more phone calls from people in the area who found injured animals – stoats, owls, kestrels – on the roads and wondered how they could help. She took them in, and worked closely with a local vet, Brian Coles, who performed intricate and innovative surgery on broken bird wings. As the title of her second book emphasised, Jane passionately believed in the intrinsically wild character of animals and vowed always to return them to the countryside.

In 1970, two unremarkable-sounding events occurred within twenty-four hours that changed Jane's life. She visited London for the annual general meeting of the WI, and she took in her first badger cub. Four months old and the size of an underfed rabbit, the petrified orphan was a girl whose two brothers had already died; she was sick with diarrhoea and suffering a horrible skin disease. Jane's account of how she collected the ailing cub from an RSPCA centre in the Midlands and nursed it back to health is matter-of-fact and unsentimental, two of Grandma's most obvious characteristics. What is also apparent is the astonishing energy she poured into this helpless creature, as if she had been waiting for years for this nurturing opportunity.

When she returned to Deer Close, her Lake District home so named because roe deer were found just over the drystone wall separating the garden from an adjacent wood, she put the cub in a wire-fronted wooden box lined with newspapers and blankets and placed it in a dark corner of her garage. She gave the cub milk-and-glucose baby food but it continued to lose weight, so she tried different foods until she hit upon a winning blend of puppy meal, minced raw meat and halibut liver oil capsules. Soon, the badger, which Jane christened Bodger in a rare moment of sentimental weakness, was accepting food from her hand without biting. At mealtimes, she would carefully brush Bodger's back with a washing-up brush. This was not to tame it but to ensure that the animal was unperturbed by contact during feeding so she could later treat its scaly skin disease. Finally, the badger relaxed enough to accept Jane painting its back with a thick and oily (but crucially odourless, given a badger's acute sense of smell) skin cream, which cured the complaint. This was typical of the imagination she applied to the nursing of animals: on another occasion she made a pair of special bootees for a great crested grebe to keep its feet wet while it was cared for – because if they dried out this elegant waterbird would have difficulty returning to the wild.

Jane watched over Bodger and anticipated her needs to an almost neurotic degree. She set humane traps in the garden for small rodents and fed the little sow live mice and voles so that she would learn to catch her own protein. As Jane's wildlife refuge expanded, she bred mice, which delighted my sister and me when we were children, especially when we were allowed to take four home with us as pets. Giving us pairs of mice was typical of Grandma; she was a great believer in animals in captivity having company. She went to considerable lengths to acquire a boar badger to stop Bodger getting lonely during her rehabilitation. Unfortunately the boar quickly escaped the artificial sett that Teddy built in the garden and disappeared, a rare occasion when Grandma's dedication to the tiny details of animal care did not end successfully.

*

The day before Grandma began looking after Bodger, she made the life-changing visit to London. Some months earlier, Jane's campaigning to stop digging and baiting had led her to submit a resolution about badgers to her local branch of the Women's Institute. This resolution proved so popular it was one of seven selected for debate at the WI's AGM. For a shy, rather diffident person, travelling to the Royal Albert Hall, standing before 7,000 delegates and persuading them of the need to give badgers legal protection was a terrifying prospect. Her family was amazed she did it but Teddy drove her there and she held her nerve. Her motion, 'That the Women's Institute Movement urges legislation to prohibit any killing of badgers other than under licence issued in circumstances where it is proved that they are causing material damage to agricultural or horticultural interests', was passed by an almost unanimous vote. Grandma's words are remarkably similar to the law that was eventually brought in, and the legislation that still protects badgers today.

In those days, the WI was an influential organisation, representing more than half a million country women. In the weeks and months after Jane's triumph at the AGM, she was emboldened in the moments when she was not nursing Bodger to give talks to other WI groups, urging members to lobby their MPs for a Private Member's Bill to protect badgers. She wrote articles and appeared on television and radio, exhorting the government to act. Photographs of badgers and other wildlife taken by her and Teddy, an enthusiastic photographer like many of the new naturalists, were exhibited in London. These efforts mingled with those of a band of badger watchers, including influential experts such as Ernest Neal, who met with politicians at the Home Office (there was no department for environmental matters then) to persuade the government to take action.

In 1973, this lobbying finally paid off and Parliament produced two Private Members' Bills offering to protect the badger. One bill, devised by Lord 'Boofy' Arran, whose pet badger would sit on his lap while he took afternoon tea, was judged by Neal to be badly drafted but comprehensive; the other, by Peter Hardy, a Yorkshire MP and keen badger watcher, was a very modest measure that sought to thwart digging by means of the laws of trespass. Arran's bill was too ambitious, thought most badger lovers; Hardy's, too limited. In fact, Hardy's was worse than no law at all, according to Jane. 'I felt so strongly that despite all my endeavours over a great many years, I would rather have seen Mr Peter Hardy's Bill dropped than go forward unamended,' she wrote.

The vast majority of Private Members' Bills never get near the statute book. But after numerous meetings Hardy was persuaded to drop his bill, leaving the way free for Lord Arran's, which was debated in Parliament. At one stage, Lord George Brown rose to remonstrate with Lord Burton, who was trying unsuccessfully to exclude Scotland from the bill. Burton, said Brown, was an old badger who should go back to his sett.

On the very last day of the parliamentary session, to widespread relief among naturalists and badger lovers, the bill was squeezed through the Commons. The Badgers Act of 1973 was a pioneering piece of legislation. It was the first time in British history that a land mammal had been given specific protection from persecution (grey seals were protected in 1914). Because it made it an offence to 'cruelly ill-treat' any badger, digging and baiting were finally outlawed, as was the use of tongs and dogs. The sale and possession of live badgers and badger pelts were also made illegal. Badgers could only be killed by 'authorised persons' who could prove their intervention was necessary to prevent serious damage to crops, poultry or property, or prevent the spread of disease.

The new law received deserved acclaim. For Lord Arran, who also drafted the 1967 Sexual Offences Act, which liberalised the laws on homosexuality, it was one of two moments that defined his parliamentary career. He later summarised them as attempting to prevent people from 'badgering buggers and buggering badgers'. The Act was far from perfect and the category of 'authorised persons' was, badger watchers feared, open to abuse. But it created a model for other wildlife-protection laws, set in motion the first serious attempts by police to tackle this new category of 'wildlife crime' and, perhaps most significantly, sent a clear message that the authorities would no longer turn a blind eye to the killing of badgers for fun.

Throughout my childhood, Grandma continued to stay up past midnight and rise at dawn to care for her wild animals. Like any kind of nursing, this physically demanding work permitted no undue sentiment; but Jane, sturdy and indefatigable, took unusually great delight in the company of her first badger, Bodger, who was not unlike a stout-legged dog, a loyal animal who repaid her tenderness with whickering conversation and vigorous play.

A growing badger has inexhaustible vitality, and Jane would take Bodger for walks in the woods around their home, the short-sighted beast trotting at her heels, gradually becoming familiar with the world beyond Deer Close. She wrote a vivid account of the little badger's instinctive alarm posture, which Bodger displayed one evening in the garden: 'She was galloping about, then suddenly stopped dead with a loud snort, and on four stiff legs puffed out her long coat as a porcupine draws forward its quills. She looked like an enormous grey puff-ball with a small black and white striped head protruding from one end.'

Despite the growing bond between woman and beast, Grandma knew she must let Bodger roam free. In the wall that separated their garden from the wood, Teddy constructed a badger gate according to Forestry Commission specifications. Now Bodger was familiar with her surroundings, she could enter the wild world. Like an anxious mother whose teenager has just left home, Jane could not quite let go. Teddy rigged up an electrical contact on the badger gate which rang a bell on the stairs of their house. This enabled Jane to wake up and precisely record Bodger's arrival and departure times from Deer Close. During the day, Jane would press a stethoscope onto the lid of the artificial sett dug into their garden and listen for the sound of breathing; if Bodger was not there she would sit on a large rock outside the second sett the young badger had dug for herself in the wood, waiting for her to emerge in the evening. 'To watch her first testing the wind for danger, coming out most cautiously, gave me great pleasure, knowing that she was taking on the inherent caution of the wild and, for all her time of contact with a human being, she could behave like her truly wild relations,' she wrote.

Jane reduced Bodger's dependence on her by supplying her only with bread and diluted syrup, so the badger would have to forage for her own protein. But Jane still seemed dependent on Bodger. Surreptitiously, she sought out signs of her. She spread sand across the various gateways to see if Bodger's prints would betray the direction she was headed. After a six-week spell without contact, the bell on the badger gate sounded and Grandma rushed into the garden. You could sense her relief when she wrote that, after a startled moment, Bodger recognised her voice and musked her boots to resume their relationship.

'We are by nature selfish and would, deep down, feel honoured if they stayed and responded with affection,' she wrote in *Through the Badger Gate*. 'To put the badger first and be completely unselfish about returning it to the wild, parting with the creature to whom one has given one's all, both day and night for so long a period, takes strength of heart and courage.' I suspect that Grandma struggled to maintain this courage more than she let on. Looking back on her account now, I wonder what was missing in her life and what absence one small badger so eloquently filled.

Young children are famously accepting and Jane, for me, was a thrilling grandma, with her smelly garage full of dead chicks, her aviaries stocked with owls, and her slideshows and speeches. I remember attending the launch of the paperback of *Fly High, Run Free*, in the days when publishers threw parties in country hotels. She stood on the terrace and gave a talk. My memory may be playing tricks but I think she had one of her recuperating owls – temporarily unfree – perched on her arm. My sister and I were as exotic as badgers at such a grown-up event, and we were talked down to by ladies in summer dresses before we escaped, as wild things should, and ran off across the springy, mossy lawn. We each got copies of her book, lovingly dedicated to us. Only later did it dawn on me that Grandma was not interested in worlds outside her own, and never offered her children or her grandchildren the attentiveness and affection she bestowed on her birds of prey and the badgers she gave names such as Baloo, Puffles, Buboo and Tiny-Toes before returning them to the wild.

Virginia McKenna, the actress turned wildlife campaigner, wrote an astute foreword to *Fly High, Run Free*. 'Perhaps the thing which strikes one most about this book is the author's lack of self – an unusual characteristic of first-person, real-life sagas,' she wrote. 'All her energies, mental and physical, are directed towards the animals and their wellbeing. She is a woman of action.' Before she entered the long twilight of Alzheimer's in the 1990s, Grandma tore through the years more intensely than ever, assisting a parade of other injured and needy badgers, giving talks about her work and campaigning against all kinds of cruelty towards animals. She was certainly not one for inaction or reflection, and there was something of a child in her instinctive empathy for small things and the prodigious energy she dedicated to their cause.

*

Not even my grandma's attention to detail could save her first badger, however. Bodger left her setts at Deer Close and in the wood and dug new lodgings in rough pasture on a nearby fell-side, which Jane eventually found after weeks scouring the bracken-covered crannies and crevices in her neighbourhood. Then, in the middle of her first winter, Bodger disappeared. The badger gate no longer swung, the bell no longer tinkled and there were no fresh tracks in the woods. The bracken Jane spread over the setts' entrances remained untouched; there were no signs of any badger digging for earthworms on the fells. Bodger was missing. 'She has never known the joys of spring and the soft night air of May,' Jane lamented when spring arrived. 'She endured a ghastly painful death some months ago.'

Although she had no evidence, Jane was convinced that Bodger had died in agony after accidentally consuming strychnine that the local farmer's mole catcher had placed in baits made from the badger's favourite food, earthworms. The farmer wanted to rid the paddock of the menace of molehills; if he had left it to nature, the badger would have probably done it for him. 'Man had once more won the endless battle he wages between his worldly betterment and the existence of these wild creatures who share our lives upon this earth,' wrote Jane gloomily. 'Although I mourn her passing, one life has been made the richer and the fuller for her company.'

In spite of the loss of Bodger, the outlook for the badgers of Britain was now encouraging. At the beginning of the 1970s, badger baiting was openly practised in many communities and badger hams were still found on pub menus. With the 1973 Badgers Act, the law caught up with the pro-badger views expressed in living rooms across the land. It would be broken on many occasions in subsequent years but the Act had one unequivocal effect: digging, baiting and other persecution declined and Britain's badger population began to rise again.

By the time this protection reached the statute book, however, the battle over badgers had already moved on. The species was now placed in direct conflict with the interests of farmers and consumers. The truce between man and badger was threatened by a new and unexpected turn of events: disease.

7

Vermin

Do you know the real truth? people whispered when I mentioned I was writing about badgers. You know those carcasses on the roadside? They were not hit by cars – they were shot in the head by farmers. Whatever the truth of this urban myth, its popularity reflected a growing awareness that there were country people with good cause to hate badgers.

'I don't like the things. I think they are horrible creatures,' one farmer told me. 'It is one of the most protected animals in the northern hemisphere and it has no enemies apart from the motor car,' shrugged a despairing dairy farmer in the Cotswolds. 'Nobody sees them as vermin. We only see a rat as vermin,' said a farmer's wife. 'Our son says the problem was that Walt Disney made the animals talk,' said another.

'*The Wind in the Willows* has a lot to answer for. And *Watership Down*. And *Bambi*,' sighed Oliver Edwards, his sturdy hands resting combatively on his broad kitchen table.

'Nobody could cry more than you watching a Lassie film,' laughed Jill, Oliver's wife, leaning on the Aga as a cat slept in the old kitchen fireplace.

'I hate it when my animals go off to the abattoir to be killed,' nodded Oliver. He looked dusty, as if he had just emerged from a shed full of straw. 'But they've had a nice life.'

I drove to Somerset on a bright spring day to meet some of those who farmed in a hotspot for bovine TB. Oliver's farmhouse was old and made of stone and sat at the end of a single-track road in the bottom of a steep valley. He was a third-generation Exmoor hill farmer with 600 acres of rough ground on which he tended 250 ewes and eighty Aberdeen Angus cows. Like most farmers these days, he could not thrive on this alone and so he and Jill had converted some old labourers' cottages into holiday homes and offered B&B too. This weekend, the farm was hosting a wedding and the bride's parents scurried through the yard worrying about the weather as Oliver sold a box of prime cuts to a man in a van from Colchester. The price was £140 but the man only had a bank card on him; send me the cheque in the post, said Oliver, trustingly. With his greying curly hair and impish smile, he must have been a charmer when he was younger and wooing Jill, his sister's teacher-training friend, who was originally from Essex. They were still a handsome couple.

Oliver's farm was currently 'shut down' because of bovine TB, perhaps the most intractable animal health problem we face in Britain. An infectious bacterial disease affecting man and other warm-blooded mammals, tuberculosis enters the body through the nose or mouth and the bacilli join the blood, which transports them to the vital organs – the lungs, kidneys, liver and intestines. Pale, spherical tubercles, or lesions, form within these organs and slowly and silently kill. TB is very difficult to detect in the early stages; rare clinical signs may include lethargy, emaciation, fever and pneumonia with a chronic cough. For centuries, it was a very human tragedy, and it still kills well over a million people around the world each year.

At some point following the domestication of cattle *Mycobacterium tuberculosis* jumped species, mutating into *Mycobacterium bovis*, a cattle-based form of the disease. In the 1930s, one-third of cows in Europe suffered from bovine TB. At an Edinburgh abattoir in 1937, 44 per cent of cattle were infected. In Britain, milk yields slumped and cattle died prematurely. *Mycobacterium bovis* was also a zoonotic disease – it could jump back to humans – and at its height in Britain in the early twentieth century, there were 50,000 human cases of bovine TB; 5 per cent of those infected died. Through breathing and via urine and faeces, cattle could pass bovine TB to other animals, including deer, cats, pigs, sheep – and badgers, who often shared paddocks with cattle and flipped cowpats in search of beetles. The bacteria could also linger in water, animal feed and soil. Despite the existence of TB in cattle across much of Europe where the badger resides, it is only in Britain and the Republic of Ireland that such strong links have been drawn between the two in the last few decades.

Being shut down, for Oliver Edwards, meant that he could still drive to and fro in his mud-splattered Land Rover, buy and sell sheep and keep any remaining healthy cattle on his land, but he was not permitted to move any live cows on or off the farm. He had been shut down because several of his herd had reacted to the bovine TB test, which is carried out annually on most farms in Britain. Oliver's herd would now be tested every sixty days. If more cattle 'reacted' to the TB test, they would be taken away for slaughter. He would need two completely clear tests for every cow before the restrictions would be lifted. Few farmers escaped 'shutdown' in the minimum 120 days, however. It was a difficult disease to eradicate and nearly a quarter of cattle farms in the South-West were shut down in 2010; in Somerset and Gloucestershire, the proportion was higher.

'It's a real shit actually,' said Oliver, lightly enough.

Oliver had always run a closed herd: even in normal times, he did not buy in any animals but bred his own replacement heifers and his own bulls. In this remote spot, his herd did not come into contact with any other cattle. 'The only animal they come into contact with that carries bovine TB is the badger. You can protect yourself – your sheds, your farm buildings. I do the best I possibly can to combat bovine TB but the thing that will travel between my farm and the neighbour's farm is the black-and-white and I can't do anything about it. Legally.' He looked at me meaningfully.

The shutdown inhibits a farm's productivity and causes its costs to soar. Like every farmer, Oliver ran his business on a neat timetable. His cows calved in March or April and the calves stayed with their mothers until November, when they were weaned and sold to a farmer who 'finished' them – fattening them up for slaughter. Winter is by far the most expensive time for a farmer: cattle must be kept indoors, often in a heated building, and usually bought extra feed and straw. Oliver's system relied on keeping as few cattle as possible in the expensive winter months. Once he was shut down, however, he had to keep the previous year's calves and feed them through December, January and February. Now, by spring, he had these unwanted yearlings as well as his usual heifers and their new calves: 210 cattle instead of the seventy he had budgeted for. 'The stress of being under TB restriction for the family, your bank balance, the stress of trying to manage your business, is horrendous,' said Oliver. 'For some farmers it gets to breaking point.'

Bovine TB cost the taxpayer £500 million in the ten years to 2011. The money was mainly spent on compensation for farmers of diseased cattle: 37,753 were slaughtered in England and Wales in 2012 because of TB. In the following ten years, it would cost £1 billion, according to number-crunchers at the Department for Environment, Food and Rural Affairs (Defra). But the compensation Oliver received for each slaughtered cow failed to cover the extra production costs of being shut down; nor did it match the real value of an animal. Compensation in England was based on the crude average market value of a dairy or beef cow. There was no valuation of individuals (this fairer system was once used in Wales, but led to fraud) and so no distinction was made between a scrawny hill breed and a top-quality bullock; Oliver's prime beasts might be worth £1,000 each but he would not receive that.

Disease is no respecter of prize cattle. 'I try to breed certain bulls to certain cows and it all goes tits up. Your best cow goes under TB restrictions,' said Oliver.

Jill pointed out it could be genuinely heartbreaking when unique bloodlines, carefully nurtured by three generations of farmer, are wiped out. 'You might just think, oh it's just an animal to produce for your living, like a pair of shoes, but it's not like that,' she said. 'It's all those years of breeding. It doesn't happen by magic. It takes years of good stockmanship.'

'Imagine,' said Oliver to me, 'that you had taken several years to write this book on your laptop. You then have to press one of two buttons. If you press the wrong one you lose all your work. The bovine TB test is like that. It's totally out of your control. You can do nothing about it.' He would keep a good calf for two years, put it to the bull, and it would deliver a calf at three. By selling that calf when it was two, he would receive his first income from those animals – after five years. With one test, bovine TB could wipe out five years' work.

The TB test was not good for farmers' sanity. 'A couple of weeks before you're testing cattle some farmers are not nice to live with because they're worried,' nodded Oliver.

'I can vouch for that,' chipped in Jill from the back of the kitchen.

'It's the stress for us, it's the stress for the animals. It's not fair on them,' Oliver added.

Before every test, the cows must be herded into a race – a railed passageway – and run through the crush, a holding department where the neck is secured by a frame so the animal can't jump back and forth. Here, a vet measures the thickness of the cow's skin as the first part of the tuberculin, or 'skin', test. The animals are then given two jabs, one of live TB and another of avian TB, the control. The young cattle are twitchy and nervous. Sometimes an animal slips and breaks its leg. If the heifers have just conceived 'you have to be very gentle – otherwise they'll lose the foetus,' Oliver said. Three days later, the vet returns and measures the skin with callipers where the cow was injected. If both lumps are the same, the cow is OK. If the live TB lump is more than 4mm larger than the control, the cow is a 'reactor' – it has reacted to the TB test and may have the disease. This is deemed either conclusive, in which case the cow is taken away and slaughtered, or inconclusive. If it is inconclusive, the animal is isolated and tested again in two weeks' time. If the test is inconclusive again, a more accurate blood test may be taken, or the cow is slaughtered.

Confusingly, even when a cow is judged to be a reactor and is slaughtered, conclusive lab tests on the dead animal regularly find it never had TB in the first place. These are 'false positives'. More alarmingly for farmers who assume that when they are given the all-clear by the vet their herd is free of TB, there are also 'false negatives' – animals that are riddled with TB but never react to the skin test. In fact, the skin test is so inaccurate that it fails to correctly detect an infected cow in about 20 per cent of cases. These undetected carriers of the disease may be responsible for many cases of cow-to-cow transmission.

It sounded like Russian roulette, with a malfunctioning gun. If you were clear at the end of the testing process, said Oliver, you hugged the vet.

How did badgers bumble into this unhappy picture? Human deaths from bovine TB in Britain were virtually eradicated with the pasteurisation of milk in the 1930s and 40s. The next step was to reduce the transmission from cow to cow. From the mid-1930s farmers were encouraged to test their herds for the disease. Compulsory testing was introduced in the 1950s, as were the slaughter of infected animals and the imposition of movement restrictions to prevent untested or infected herds from being transported to market. In response to these measures, the disease steadily declined in the postwar years, when farmers also prospered more generally.

Then, in the summer of 1971, a badger was found dead on a farm in Thornbury, Gloucestershire. This was an ordinary enough occurrence: badgers were common in the neighbourhood. For some years, however, cattle in the area had suffered particularly badly from bovine TB. Local farmers pointed a finger of suspicion at *Meles meles*. Could badgers be spreading the disease?

The carcass was tested by Ministry of Agriculture vets. They identified *Mycobacterium bovis* in the body. 'A sinister cloud appeared over the world of badgers,' wrote Ernest Neal, who was 'dismayed' that it swiftly became 'a political issue'. The possible link between badgers and TB was splashed in the press and appeared as a storyline on the BBC Radio 4 soap *The Archers*. Post-mortems on dead badgers found near Thornbury identified three more as carriers, although most of the tested carcasses were TB-free.

The history of badgers and bovine TB is full of bitter ironies, from the fact that it was cows who originally passed the disease to badgers (and continue to infect badgers) to the strange quirk of chance that the disease should have materialised in badgers just months before they finally gained legal protection. 'The discovery of this first dead badger indirectly put badgers under a much greater threat than before the passing of the Badgers Act,' my grandma wrote in 1979. Like Neal, she believed 'the whole affair was blown up out of all proportion'. At the same time, the broadcaster and countryman Phil Drabble argued that cattle had become more susceptible to diseases such as bovine TB because selective breeding aimed at producing higher milk yields or rapid growth gave birth to less robust animals. 'Nature concedes nothing without exacting her price,' wrote Drabble in the 1970s.

Without conclusive scientific proof of badgers transmitting the disease to cattle, the Ministry of Agriculture decided that *Meles meles* in bovine TB hotspots should be killed. 'The Ministry, grasping at the straw, started a witch hunt,' declared Drabble. There was one problem: the 1973 Badgers Act. Snaring badgers could leave officials liable for prosecution under the new law, while gassing setts would fall foul of the 1911 Protection of Animals Act. Ministry of Agriculture officials suggested that farmers could be authorised under the Act to kill badgers and, in 1974, staged a demonstration in the Cotswolds to teach farmers the most effective way of snaring them. This caused public outrage and the Ministry hurriedly cancelled other badger-murdering demonstrations. Badger lover Ruth Murray brought a private prosecution under the Act against the Agriculture Minister and his staff for killing the animals. The prosecution was unsuccessful – that badgers could be legally killed to 'prevent disease' was an important caveat in the Act – but the case established that self-locking snares were inhumane in the eyes of the law.

Undeterred, the government amended the law in 1975 to permit the gassing of setts under licence. The first culls began of badgers living close to farms where cattle had contracted bovine TB. A white powder called Cymag was placed in the entrance to a sett, producing hydrogen cyanide, a poisonous gas, on contact with moisture. This spread through the sett and, supposedly, put badgers to sleep. In the 1970s, an eradication programme was undertaken at Thornbury, where the first diseased badgers had been diagnosed. Before 1976, thirty-nine herds of cattle were found to be infected with bovine TB in an area of forty square miles. Conveniently, this area was almost surrounded by so-called 'hard' boundaries for badgers – the Severn Estuary and the M5 – so they could be gassed without their vacant territories being instantly recolonised by newcomers. The last cow with TB was found a few months after the gassing concluded, in 1977.

Thornbury, the first serious government-organised badger cull, recorded the following decrease of TB in cattle: 100 per cent. This success has been cited by farmers as an example of how dramatically a draconian badger cull could reduce the disease in cattle. Unfortunately for the farmers, though, wiping out the badger population across much larger tracts of the West Country and Wales would enrage most of the people living there and probably contravene the Bern Convention, a European treaty obliging Britain to conserve its native wildlife. Any badgers that recolonised such cull zones would also be genetically disadvantaged, lacking any immunity to bovine TB from natural selection. 'Frankly, people who advocate a Thornbury approach need their heads testing,' judged Chris Cheeseman, the biologist who set up the government's badger research station at Woodchester Park in the Cotswolds and studied the disease in badgers for more than twenty-five years.

By 1982, hydrogen cyanide had been pumped into 4,000 setts. In less than a decade, the killing of badgers had moved from being an illicit private pleasure to, apparently, an official necessity. Those who eradicated badgers were no longer disreputable members of a dog-fighting underclass but a new breed of university-educated vets, bureaucrats and scientists. 'Gassing carries with it thoughts of mass destruction and of unselective killing of defenceless victims. In itself it is abhorrent to many people; in the hands of officialdom it is potentially explosive,' wrote Robert Howard in *Badgers without Bias* in 1981. Popular opposition quickly grew to the government gassing of badgers. Before the general election in 1979 the *Western Daily Press* ran a 'Save Our Badgers' campaign (a marked contrast to its pro-cull position in 2012). The Conservative government introduced a moratorium on gassing which became a permanent ban in 1982. Officials were spooked by scientific experiments conducted in shadowy

Ministry of Defence establishments that showed the failure of gas to cleanly kill the residents of the labyrinthine setts, leaving many badgers to suffer a lingering, agonising death.

Since 1971, the clamour to kill badgers has waxed and waned with outbreaks and retreats of the disease, and the controversy over badgers and bovine TB died away during the 1980s. Badgers continued to be killed under licence if found on farms where the cattle were infected. But these were trapped live and then shot, which was considered more humane than gassing or simply shooting them because it was difficult to administer a clean kill to a mobile, thick-hided animal in the dark. There was little argument, however, because so few badgers or cattle were killed. By 1981, just 0.5 per cent of cattle in England and Wales reacted positively when tested for the disease, thanks largely to much stricter controls on their testing and movement.

During the 1980s, 90s and 2000s, the badger population steadily increased once more. Police officers trained in wildlife crime actively sought to stop digging and baiting, and participants were prosecuted. In 1992, badgers were further protected by a new law that criminalised any interference with setts. Only with a government licence could a sett be moved or 'closed' by developers or farmers, and that could only be undertaken after a cuddly new breed of badger consultants ensured no harm came to the residents.

Oddly, for all the impressive badger science, there is a paucity of accurate data about their population growth in recent decades. Predictably enough, badger lovers downplay talk of a surging population, stressing that badgers don't breed like rabbits, huge setts may contain fewer than ten individuals, and it is estimated that 50,000 – approximately one-fifth of the adult badger population – are killed on the roads each year. Equally predictably, farmers are convinced that the badger population is out of control. Michael Dougdale, a conservation-minded farmer, took me around his land in Shropshire, where there have always been badgers. When he arrived in 1968 there was one sett. Now there were twenty-two and the older setts had been expanded beyond all recognition. 'People still say they are rare,' said Dougdale, exasperated at these badgers' predation of ground-nesting birds and hedgehogs. 'They're as common as muck.'

In spite of the difficulties I had in spotting badgers, farmers' anecdotal accounts of a significant rise in numbers rang truer to me than the claims of badger lovers. The MP Peter Hardy estimated there were 40,000 badgers in Britain when the Badgers Act became law. The last scientific survey, undertaken between 1994 and 1997, more accurately calculated that there were 50,241 badger social groups in Britain. If an average group had five animals, there would be 250,000 badgers in Britain; if it had eight, there would be more than 400,000. Most remarkably, the survey estimated an increase in the badger population of 77 per cent since the previous study in 1988.

Defra is currently funding a long-overdue survey to obtain up-to-date population statistics – but what did scientists estimate the population to be in 2012? 'My guess is it's increased by quite a lot but I wouldn't want to say by how much,' the badger scientist Tim Roper told me. The decline of persecution was the big factor in their population growth but urban badgers have also increased because, thought Roper, people fed them more. In Wytham, Chris Newman and Christina Buesching argued that reduced persecution from humans was not the prime driver of a population increase: in their woods, badgers have been undisturbed since the war but the population only began to rise significantly in the 1990s. In southern Finland, where badgers are hunted as they always have been, numbers have doubled in the past fifteen years. They are moving north, and are now inside the Arctic Circle. Newman was convinced population increases are caused by climate change. Milder winters have given badgers more foraging opportunities. Elderly badgers survive to breed another year; females are in better condition and more of their litter are likely to survive each spring.

A final, ironic cause of the badger's population rise is the farmers themselves. Modern agriculture is inadvertently cultivating *Meles meles*. Drained land enables badgers to dig setts where once they would flood. Summer droughts used to kill many cubs but the current ubiquity of rape and maize carries them through lean times: maize ripens at a perfect moment, in autumn, when badgers must fatten up if they are to survive the winter.

If you placed two graphs side by side showing the estimated increase in badgers since the 1980s and the rise of TB in cattle, you would see two ascending lines. This is far from scientific proof of causality but many farmers believe the two trends are linked.

'It's just unfortunate that the badger numbers have increased so dramatically since the badger was protected in the 1970s,' said Oliver Edwards over the kitchen table in Exmoor. 'Take away the bovine TB and you see an increase in a population that you can't control, that as countrymen we used to control. If something gets out of hand it preys on things – hedgehogs, ground-nesting birds, bumblebees. Disease is rife in some of the badger setts because they are totally overpopulated. The badger is a native to the UK, it's part of the countryside. It's such a shame it's suffering as well. If the numbers were reduced we'd have a lot of healthy badgers.'

Most farmers are in no doubt that badgers spread bovine TB to cattle, and that a rise in the badger population is behind the resurgence of the disease. My grandma and the badger lovers of the 1970s believed the creature was a scapegoat, and many badger enthusiasts, including the Badger Trust today, still argue that its role in transmitting bovine TB is vastly overstated. Who is right? The farmers' first conviction is supported by science but their second is not. Long-term data gathered by Chris Cheeseman's team at Woodchester Park shows that there is no link between the density of badgers and prevalence of TB: density increased, plateaued and then fell while the incidence of TB fluctuated without any correlation.

The animal most likely to pass TB to a cow is another cow – the disease has struck cows on the Isle of Man, where there are no badgers. Nonetheless, the scientific evidence that badgers are the most probable 'wild' source of TB in cattle is well established. Lord John Krebs, a distinguished zoologist after whom the laboratories outside the badger heaven of Wytham Woods are named, led an independent review of the evidence of bovine TB in cattle and badgers in 1996. He judged that the sum of evidence strongly supported the view that badgers were a 'significant' source of infection in cattle, although he conceded that 'most of this evidence is indirect, consisting of correlations rather than demonstrations of cause and effect'. This is still the case today.

The clearest direct evidence of the transmission of bovine TB between cattle and badgers comes from a 2012 genetic study of bacteria from twenty-six cows and four badgers in Northern Ireland. Scientists traced mutations in the bovine TB bacteria as it passed from animal to animal and found 'indistinguishable' bacterial types in badgers and cattle from nearby farms. Although the research did not establish the direction of infection – whether it was spread from cattle to badgers or badgers to cattle – scientists are certain that badgers transmit the infection to cattle as well as vice versa. 'There is no doubt in my mind that badgers are implicated in the incidence of TB in cattle,' said Chris Cheeseman. 'If you remove badgers you can reduce the prevalence of TB in cattle.'

Bovine TB is understood to exist in a 'wildlife reservoir' of wild animals that regularly reinfect cattle, because even when it is eradicated from herds, it still recurs, and the disease continues to be found in geographical hotspots, mostly in the West Country. Research has identified different genetic strains, or spoligotypes, of bovine TB, and even after some decades these different strains remain associated with specific geographical areas – a certain TB strain is mostly found in Devon; another is concentrated in Gloucestershire; another in

Cornwall. If bovine TB were only passed from cow to cow, the spoligotypes of the disease would quickly become mixed together by modern agriculture's busy trading of cattle, and it would also be more evenly distributed across the cattle population in areas such as Scotland and Yorkshire, where it is almost unknown.

The wildlife-reservoir theory is supported by experience overseas: possums are a reservoir of bovine TB in New Zealand; bison in Canada. Bovine TB may occur in many wild animals but scientific studies have found that none are as likely to spread it to cattle as decisively as badgers, which combine relatively high rates of infectiousness with ample opportunity for encountering cows in fields. The most comprehensive review of TB in wild mammals found that only red and fallow deer in theory posed as great a threat as badgers – combining high rates of infection with the possibility of encountering cattle in the countryside. But in practice these deer have not been numerous or widespread enough in the past to explain the geographical distribution of bovine TB in cattle.

For Chris Cheeseman and other scientists, however, a key question remained: *how much* of the TB that occurs in cattle is due to badgers? Despite decades of study in many different branches of academia, from veterinary science to epidemiology, scientists have not clearly established the answer. 'The leading experts have a view – you can take an average and it is about 30 per cent,' said Cheeseman. Tim Roper agreed but, with a wry grimace, said that the best scientific estimate was incredibly broad: 'Anything between 15 and 75 per cent.'

In 1998, the last year for which data was collected, 23 per cent of badger carcasses tested were found to have bovine TB. But not all infected badgers will be infectious. The best way to demonstrate the disease's transmission scientifically would be to develop a pathogen in a lab and see how it spread. As Roper put it: 'If this were mice you could do an experiment with 200 of them in a lab and get an answer in a couple of weeks.' But cows and badgers are not good lab rats.

The biggest scientific endeavour to examine badgers and bovine TB began in 1998 with the Randomised Badger Culling Trial, as recommended by Lord Krebs. This seven-year government-funded operation culled 10,979 badgers, at a cost of £49 million, in an effort to determine whether killing badgers reduced bovine TB in cattle. Unlike much smaller, unscientific culls, the RBCT did not provide compelling evidence that killing badgers reduced TB in cattle. It found that, at most, culling could reduce the incidence of TB in cattle by 12–16 per cent over nine years. The Labour government of the day heeded the recommendation of John Bourne, the scientist who led the trial, that culling badgers could make 'no meaningful contribution' to the control of bovine TB. Then, confusingly, the coalition that replaced Labour in 2010 studied data from the same trial and took the opposite view. Once again, badgers would be culled in our countryside.

In the months before this supposedly 'farmer-led' cull was due to start in bovine TB hotspots in Somerset and Gloucestershire, most farmers I spoke to said their ideal solution would be if they were permitted to control the badgers themselves. Evan Thomas, a retired dairy farmer from Carmarthenshire who had witnessed the rise, fall and rise again of bovine TB, was typical. If a farmer was hit by TB, he argued, that individual should be allowed to do what is necessary to control badgers on his land – gas the setts, or snare (using the snare deployed by Irish badger culls which enables non-target species to be freed) or shoot individuals, just as farmers still control foxes. Another farmer put it very simply. 'Take the blimmin' [legal] protection off,' he said with feeling. 'It wouldn't cost the government any money and farmers wouldn't go out and shoot badgers all night because we haven't got time and we're too bloody tired in the evening.'

Not all farmers felt the same. Of course there were plenty of what they disparagingly called 'hobby farmers', city dwellers turned good-lifers, who didn't like the cull. But some 'proper' farmers did not want to kill badgers either.

David and Patsy Mallet tended 160 acres on the edge of Dartmoor in the heart of the bovine TB country, managing a beef suckler herd of eighty cows with two bulls, and 300 breeding ewes. Like many farmers I met, their life seemed both more satisfying and more gruelling than that of an average person: they had an airy, barn-like farmhouse with a beautiful kitchen and views of the moor from all their windows; they had space, fresh air and they worked at home; and yet their working lives were far more relentless than the careers of those of us who don't labour on the land.

David, a third-generation farmer, was short and fit and dashed in and out in his red boiler suit, tending to the ewes and their new-borns. Lambing was the busiest time of year and the Mallets lambed

151

outside, which was even more challenging. Patsy, a lively, forthright –
I had yet to meet any other kind of farmer's wife – blonde woman, had
come to the West Country from Cambridgeshire with Sophie, her
young daughter. One day, she saw David on a neighbour's land and
thought, Oh, who's that? They have been together ever since. Sophie
was now at university.

If you find a contented farmer, David's father used to say, the best
thing you can do is shoot him because he'll only go around upsetting
the rest. David and Patsy were not content; they, too, were currently
shut down because of TB. Despite this, they conceded, they had
been 'doing fine': China was buying a lot of British beef and market
prices were rising. Farmers are said to prosper during recession and
war.

The Mallets' experience of the tensions of the TB test matched that
of every other farmer. 'The test is a nonsense. We've had cows killed
and still not tested positive,' said Patsy. 'That's a big flaw. It's frustrat-
ing. It's immoral. We've got this cow and she's not even got TB and
she's killed and you take compensation. It's almost like blood money.
It makes a hard industry harder.' Despite the frustrations they shared
with every TB-hit farmer, however, the Mallets reached the opposite
conclusion to most. 'We've got to work with wildlife. It's very easy to
demonise one animal rather than look at your industry,' said Patsy. 'If
you're meat producing and dairy producing in a way that means great
swathes of wildlife have to be killed, that is unethical.'

Badgers lived on the Mallets' land, and damaged their pasture
when they rootled up the grass in search of leatherbacks, the larvae of

the cranefly. 'We're not opposed to the badger cull because they are "little darlings",' said Patsy. 'I wouldn't want to hold a wild badger and I wouldn't want to be a hedgehog, but I'm not demonising them. They have their life and we have ours. We have to work in partnership with them. We need to get a grip on bovine TB but in the right way.'

The Mallets were unconcerned about the apparently rising badger population. 'Like all wildlife we can live together,' said David. 'Badgers spread bovine TB but so can deer and rabbits and cows. There are shedloads of ways it can pass.' Nature regulated itself, Patsy argued; animal populations depended on food supply and farmers were the architects of their own fortunes and misfortunes. The Mallets had foxes on their land but their sheep were not troubled, she explained. Healthy ewes will protect their newborn lambs, even when they are born outside; the farmers' good husbandry – keeping an eye on their flocks – protected both. The Mallets knew farmers who could not be bothered to clear up dead sheep on the moor. Leave carcasses lying around and you soon have a fox problem. Lots of farmers shoot foxes at night; do that, and the rabbit population soars, and your pasture is nibbled to pieces. 'Lamping', the sport of shooting rabbits at night, often leaves rabbit carcasses around, and more food for the fox. 'Food supply, always,' said Patsy. 'It's a chain. I've learned that from the way David and his father farm.'

I assumed the Mallets must be in a tiny minority, but they believed plenty of farmers privately nursed doubts about shooting badgers, afraid publicly to contradict the prevailing view in their village or the received wisdom of the National Farmers' Union, which was strongly in favour of the cull. The coalition government thought the cull would be 'a vote catcher in the countryside', said Patsy, which was 'really patronising' and gave 'the signal to your daftest local person, your badger baiter', that it was OK to kill badgers. David feared it would trigger a public backlash against farmers. 'It will cost too much and the general public aren't going to support us for doing it,' he said. 'From a business point of view it's not good,' added Patsy, arguing that the cull would render British farming unethical. 'We've got some of the best welfare conditions in Europe and then we get this cull.'

Other opponents of the badger cull questioned whether bovine TB was really the disaster for farming that it was portrayed to be. The number of cows slaughtered because of it rose by 10.2 per cent in Britain in 2012 compared with 2011, although this was partly because of a 5.7 per cent increase in testing and is a tiny fraction of the 8.3 million cattle in the country. There was little risk to public health: human cases of bovine TB constitute less than 1 per cent of the total number of human TB cases in the UK. And the cattle slaughtered because of the disease were also a very small proportion of the premature deaths among Britain's cattle. A farming industry survey of 20,000 dairy cows found the main reason for culling animals in 2011 was because a cow was not in calf (24.88 per cent), followed by mastitis or high somatic cell count (an indicator of low-quality milk, 17.64 per cent). Lameness, age, udder problems, calving injuries, accidents and a general category of 'died on farm' were each the cause of more premature deaths than 'infectious disease', which constituted only 3.23 per cent of dairy cow deaths in the survey and included bovine TB and other ailments such as bovine viral diarrhoea. The British Cattle Movement Service noted in 2012 that 240,000 adult cattle die on farms each year of unknown causes.

Perhaps the most interesting thing about the cull was the fact it was so contentious. Many nations wouldn't bat an eyelid about the pragmatic control of wildlife, with guns if necessary. I wanted to ask a similar question of both sides. Why do you care so much if a small number of farmers want to shoot a relatively small number of badgers to control a disease? And why do you care so much about whether you are allowed to kill badgers when the most exhaustive scientific study found it might reduce the incidence of TB in cattle by just 12–16 per cent?

Many farmers felt the opposition to the cull came from a society that had lost touch with the real countryside. 'People living in London and Bristol don't really know the countryside, do they?' said Michael Lee, the father of Somerset farmer Nick Lee. 'People seem to think if we stop farming a field it will be marvellous,' said Evan Thomas, the old Welsh dairy farmer. 'If you leave nature alone you get an awful mess.'

Their generation was mystified by contemporary attitudes towards wildlife. 'Years ago most of the big estates were keepered and their jobs were to keep the predators down,' remembered Michael. 'There would be badgers, stoats, weasels and foxes hung up on the fence because keepers needed to protect their jobs and the boss would come along, see the animals and think, "lovely job".' In those days, the dawn chorus was deafening, there were so many songbirds. Nowadays, they've gone. 'All the predators are protected,' added Phyllis, Michael's wife, shaking her head. But the countryside these farmers remembered and their attitude towards 'vermin' were as much a product of a relatively short period of history – in which the gamekeeper ruled – as our indulgent attitude towards predators is a product of our own time. And our sense of what is real countryside, and the wild animals we permit and forbid to live alongside us, will almost certainly change again in the future.

Farmers claimed they needed to shoot badgers to control disease and protect their livelihoods. A 12–16 per cent reduction in bovine TB outbreaks created by a badger cull sounded modest but it mattered when profit margins were so tight and TB compensation did not adequately cover losses. When I talked to Chris Packham, the naturalist and broadcaster who studied badgers for five years as a student, he was sceptical about the efficacy of a cull but emphasised that farmers were under pressure. Supermarkets exerted more influence than ever on prices; hours were arduous; bovine TB was the last straw. 'The rural economy is a difficult place to make a living, and we don't help. We want cheap food and we'll buy it from overseas,' he said. 'If we all helped our farmers more – and some nations in Europe help theirs a lot more – then issues like this wouldn't be quite so critical.'

Evan Thomas took the long view. Apart from during the two world wars, he believed, no British government had cherished farmers since the repeal of the Corn Laws in 1846, when politicians chose to feed booming industrial towns cheaply over protecting farmers. The Empire reached its zenith during an era of cheap imported corn. 'That's what made us great. We weren't paying the going rate for our workers to get home-produced food because we got it cheaper from abroad,' said Thomas. And food, he argued, was still too cheap. 'One of the yardsticks of cheap food is that we can afford to waste 30 per cent of it.'

Even so, the desire to kill badgers seemed more visceral than a simple question of economics. Farmers felt politically neglected, unloved and untrusted by the authorities; stripped of their own authority and independence by government regulations and the competitive global economy. At best, their work was invisible to the rest of us; at worst they were figures of ridicule, rednecks and yokels. The cull was a rare example of a government handing back a sliver of autonomy – giving a gun to those who worked the land and saying, Do what you need to do, we trust you.

Ultimately, part of the answer to both questions – Why is the pro-badger majority so protective of a common animal? and Why are farmers so convinced that killing badgers will solve their problems? – could be found in the chasm between those who work on the land and those who don't.

Farmers see things differently. The Great Western railway line to London passes through a gorgeous green patchwork of Somerset pasture that is idly admired by many passengers. 'I take the train to London and I think, "there's not many animals around" or "oh, it's dry" or "that wheat needs a little bit of fertiliser",' said Oliver. He reminded me of a teacher in *Akenfield*, Ronald Blythe's 1969 portrait of a Suffolk village, who remarked on how the old villagers communed with nature. 'They will walk and see everything. They didn't move far so their eyes are trained to see the fine detail of a small place. They'll say, "The beans are a bit higher on the stalk this year ..."' Patsy Mallet made a similar point about our blindness to the countryside. 'Looking into a field and thinking it looks great doesn't account for the amount of work that goes into it,' she said. When those of us who don't labour on the land gaze at fields and hedgerows, we cannot really perceive what is there. The countryside is our escape, our plaything. We no longer consider the money and toil invested in healthy pasture, or the ruin of weedy fields; we don't see the cost of a crow, or a badger. It is not so much our romanticisation of rural life as our loss of the language of the fields. We admire it, and think we know it, but the countryside is as opaque to most of us as quantum physics or the global banking system.

The flight from the land and a loss of understanding of rural life have been going on for 150 years, but the farmers I spoke to felt they had become even more pronounced in the last twenty years or so. 'The French have always had this connection with land and food,' said Jill Edwards. 'Nearly all the French have cousins in the country. There doesn't seem to be this divide between urban and rural people in the way there is here.' Patsy Mallet noticed how fewer villagers worked on the land in the two decades she had been a farmer's wife. She used to make packed lunches for people who came and helped, for money, at harvest time. 'They don't want to bother now,' she said. Seasonal farm work was the preserve of Poles, Portuguese and Lithuanian labourers. 'People want the countryside parcelled up nicely for them,' reckoned Patsy. 'They think they are in the countryside because their postcode says they are.'

I was no different. The farms I visited and the kitchens I sat in were my idea of heaven. I loved the sights, sounds and smells of the countryside but I was a sensory tourist, I did not labour on the land. Nor was I trapped by it. 'My God, what a millstone around your neck,' said one farmer's wife, pitying the children who would inherit her farm. There was a gulf between me and the natural world and I felt it keenly every time I tried to write about nature. My descriptions of the natural world could never be as intimate and knowing as John Clare's in the nineteenth century, or Henry Williamson's in the twentieth, who despite being a professional writer bought a farm and absented himself from many things that kept him from the countryside. I was one of the estranged and alienated. I could not see a badger as a true countryman would.

8

Mint Humbug

After I had tried, and failed, to find the badger in strongholds such as Somerset, Shropshire, Devon and Wales, a canalside in the West Midlands sounded like a completely illogical place to seek *Meles meles*. But then a colleague from my office, Julia Kaminski, and her partner, Mick, volunteered to give up their Saturday night to take me to a desiccated gulley by a canal in Wolverhampton. Even if this seemed a long shot for what would be another weekend away from my girlfriend, Lisa, I decided expert guidance would be helpful if I were ever to see a badger in the wild.

A mate of Julia and Mick's called Andy was walking home from work along the towpath of the canal one evening the previous year when he bumped into a badger. Reasoning that there must be a sett near by, he eventually found a vast network of excavations on a ramshackle slice of land that had somehow escaped development. This was not, in fact, as improbable as it first appeared. Badgers have burrowed into the banks of canals in several places in Britain, which can be hazardous for both badger and canal. I once met a badger that had been unable to clamber out of the Grand Union Canal in Birmingham without help, and the bedraggled animal had been brought to an animal hospital where I watched it recuperating, cowering, under a pile of straw. Canalside badgers had even been immortalised in print – by Janni Howker's *Badger on the Barge*, a sensitive children's story about a young girl who befriends Miss Brady, an eccentric old lady who lives on a canal boat and keeps a badger that resembles 'a fat bear' and bounces along 'thumping the planks, making a chickering snuffling whinny, like a tiny horse'.

Colleagues usually show only their conventional side in the office so it is always fascinating to see their relaxed selves, away from work. Julia was an attractive, ageless sort of woman whose secret life embraced a quiet passion for all kinds of animals. She volunteered at an animal rescue centre and lived with Mick, a bluff, amusing Black Countryman, in a cottage hung with beautiful photographs of wild animals. Mick was a birdwatcher who claimed not to tick birds off a list but admitted to photographing fifty-three species in their garden.

We parked up by the canal just after eight. Two blackcaps duelled in the bushes as the sun lowered itself behind semis, allotments and a 1960s school with a tightly mown field. The alleged badger hotspot was a steep gully of blackthorn, hawthorn and elder thickets, overseen by larger beech trees and a gigantic ash adjacent to the canal. Little grew below their canopy and a pile of bricks and an old plastic bottle marked where teenagers had lit a fire in the dirt. A track meandered along the gully bottom, a desire line created by evening dog walkers.

It was well peopled, not a wild place, and all along the steep bank were badger holes.

We walked stealthily along the top lip of the gully in daylight, a few minutes past eight o'clock. For a moment, our strategy seemed rather vague. Mick disappeared, silently. Ten minutes later he reappeared, and led us around a blackthorn thicket where a thrush was loudly repeating three notes.

The energetic badgers of the Black Country had dug five grand entrances into the gully edge. Each hole was in the centre of its own crater, worn down by comings and goings. The bare earth was rose-coloured and scattered with wind-torn new leaves, which looked like litter on the pink earth. Luckily, the wind was blowing our scent away from the sett so we crouched down on the bank above it for what I expected would be a long vigil. There was a patter like hail. I glanced around. Mick had his hand in a bag of peanuts and was hurling them onto the badger holes below. Eileen Soper, who wrote of 'the sense of rest in the countryside' between dusk and darkness, also lured her badgers with peanuts. After a while, her scent became connected in the badgers' minds with 'the sought-after delicacy', and she fed one animal by hand.

'When you see them for the first time, you'll probably squeal,' warned Julia. 'Everyone does it.'

We were in full view but, at the top of the gully, we were also above the badgers. Sitting here with fellow watchers and a pair of excellent binoculars at a companionable hour was very different from my solitary badger missions: jollier, and less bewitching. Although it was gloomy under the trees, the sun had only just dipped down. I realised I had set out too late on many of my previous badger-watching attempts.

After five minutes, Julia pointed. We all followed her gaze. It was the movement of a leaf, a false alarm.

Cars trundled along the trunk road and two blackbirds halted their liquid evensong and began a raucous chupping. With a strut and a flap characteristic of many male animals, the duelling rivals then screeched like tropical parrots. Did badgers take heed of blackbird alarm calls? Were we being betrayed? In the bottom of the gully a dog sauntered through the trees, its unseen owner somewhere beyond, and I quietly abandoned any hope of seeing a badger; a dog – in the hands of men – was the only thing a badger here would fear.

Soon afterwards, I noticed something white, jerking like a finger-puppet, in the crater of the second hole along from us. It bobbed around for a few seconds. Then it rose up some more. It was a perfect little badger nose. I suppressed a squawk, and turned to Julia and Mick. They trained their binoculars on the hole. A bit more bobbing around and the nose was followed by two beady eyes and a slender face. Fifteen minutes after sunset, the badger brought its whole body into view, dabbing its snout on the bare earth – dab, dab, dab – searching for something.

The badger paused, and appeared to be squinting, with the short-sightedness of a nocturnal animal, in our direction. He was a magnificent boar. Two black stripes ran from the back of his neck over his small ears and elderberry eyes, turning down around the muzzle to meet his lips before his nose. These stripes were far more pronounced than I had expected, blazing out, as exotic as a zebra, as vivid as a mint humbug. His was a striking, handsome, look-at-me face; his dark-grey fur bouncy and luxuriant, as if he had just emerged from a blow-dry in a salon, and not from a dark, rather fetid hole after a long sleep. He had a curious, dainty tail, 'an amusing anticlimax to the long body' as one badger watcher put it. Badgers' tails are often described as being like a dishcloth but this one was small and perfectly formed, pointing downwards from its rear as if to say, 'I was here', which is exactly what badgers do when they go musking.

A second nose, and another perfect black-and-white head, bobbed up from the crater. This lighter-coloured, slimmer animal emerged more sinuously. The badgers who rootle through stories, poems, songs and graphic novels are noisy and blunder in the woods. This animal – the sow, I supposed – moved with a supple grace, sliding up the slope like a sophisticated miniature hovercraft. Because her short legs were tucked under her luscious fur she resembled Dougal from *The Magic Roundabout*, although her movements were much more fluid.

She followed her mate, snouting along the ground. I wondered what they were doing and then I remembered the peanuts. Systematically, they circled their sett, hoovering up every last nut. As they came closer I could hear them snuffling – fffnnufff, fffnnufff – as they pressed their delicious noses to the bare earth. After ten minutes of peanutting, a third came out; timid, slender, and a much paler grey, the colour of dry asphalt baked on a country road by many a summer. She was a yearling, one of last spring's cubs.

A dog barked, not so far away, but the badgers did not flinch. 'Hey!' a teenager shouted in the distance. Again, it caused no alarm. Watchers have recorded badgers' skill in differentiating between threatening and innocuous noises. A badger will ignore a deafening helicopter overhead but will bolt when a twig cracks under a human boot. I wondered how suburban badgers learned to evaluate the vast spectrum of strange human noises. If they didn't ignore the din of the city, they would never leave their setts.

The light was ebbing away but we could still pick out these three wild beasts perfectly. My mind struggled to process what my eyes were recording. With the rumbling traffic, the rubbish, the barking of dogs and hollering teenagers, this was a quotidian suburban scene. And yet it was every bit as dramatic and surreal as a first sight of a savannah populated by giraffes and lions. Like those animals, these badgers looked foreign. How had they lived here all along? How could there be several hundred thousand in Britain when I had never seen one like this in the wild before? This was their native land, their home. Badgers had almost certainly lived in this sett for longer than the century-old suburbs near by, enacting their own family dramas just as human populations did, only far more obtrusively, next door. These badgers did not look cowed, they did not appear under threat, they were not helpless victims; every twitch of muscle and stretch of sinew spoke of their belonging here.

Mick's leg had gone dead, and he withdrew from the increasingly crepuscular gully. I was transfixed, reluctant to stop gazing at these teeming lives unfolding beyond ordinary human perception. These badgers had their own world to attend to and it was nothing to do with us. I had an urge to interact with them, that childish and yet fundamentally human reflex to make an impact: to throw a pebble in the water, to chase a bird, to swat a fly. I wanted to see what they would do if I whistled or said hello. I wanted to communicate with them.

As I stood up silently, one badger turned and picked up my suspicious scent. I knew I must back away and disappear without them noticing me, as badgers did with us every night. Their lives were hard enough without me giving them a fright, and a fright was all I could give them; this was how relations between badgers and humans were fixed.

When I embarked on this journey into Badgerland, I had wanted to study the animal dispassionately, assess its part in spreading bovine TB carefully and give a fair hearing to those who believed it should be culled in the same way as we control many wild species. At that moment, marvelling at the grace of the badgers of the Black Country, I realised this would be harder than I anticipated. These animals were so charming and innocent; stoic, independent and enduring. In this twilight, it seemed a basic decency to give them peace; they had a right to a life free from our persecution, and from our charity.

9

Dainty, Big Un and Elaine

As dusk fell, a drop-leaf oak table was wheeled from a previously overlooked location to stand before the French windows. In the middle of the table stood a washing-up bowl filled to the brim with soaked dog biscuits, chunks of bread stewed in sausage fat, and chopped apples, pears and bananas. Next to the bowl were old four-litre and two-litre ice-cream tubs, packed with small square sandwiches made from brown and white bread. These were filled with grated peanuts. Another tub was brimming with dried peanuts and a final ice-cream container bore the ultimate delicacy: cold sausages, carefully sliced, lengthways. 'I have to cut them into thirds these days, they are so expensive,' said Judy Salisbury, laying out this spread as deliberately as a priest assembling wafers and wine for Holy Communion.

In the silence at the end of the day, the grandfather clock ticked loudly in the corner of this cold, quiet house half-buried in a Cornish hillside at the bottom of a twisty lane with grass running down its middle like a Mohican. 'I look forward to this every day,' said Judy, clutching her kettle and descending her patio steps rather gingerly to refill a water bowl. She returned, half-drew the curtains across the window that gazed down on the estuary below, and we settled back in her living-room chairs angled for the view, two strangers side by side, an elderly lady and younger man, dressed in matching blue boiler suits, contemplating the tide creeping across the mud.

Judy Salisbury's lonely house on the edge of the Camel Estuary looked over a typically British slice of countryside, containing varying degrees of wilderness, pasture and suburbia. To the left, a headland beyond which you could make out whitewashed cottages on the edge of Padstow. To the right, the estuary snaked around to well-heeled Rock. Straight ahead, over the three caged metal spans – a Forth Bridge in miniature – where the old railway crossed the water, were high cliffs, and in the gap between them the ocean began, a line on the horizon that sometimes frothed and sometimes lay flat. In the foreground, a stepped patio and a mature garden, cosseted from the worst westerlies by shrubs rounded from years of pruning, tumbled to the estuary edge. There were horses on a far field, voices in a near field, and two cyclists wobbling across the railway bridge.

Judy was a slender woman who had reached eighty and was still elegantly tall, with grey curly hair that had not yet turned white. Her home had been built as a holiday retreat by her second husband's family in the 1930s, in the days when you could buy a glorious patch of land in the middle of nowhere and construct your Dunroamin. Judy moved here when she met Robin. He died after an illness a few years ago. 'We were very happy,' she said, when I paused by a photograph of Robin sailing. 'He was a very gentle man, much like I imagine you are.'

We had only just met and Judy's trust was disarming. I had written her a letter a week earlier after I was told about a lady in north Cornwall who fed an enormous number of badgers. It sounded like Judy was both frail and not the sort of person to suffer fools, and I did not expect a reply. But she telephoned immediately upon receiving my letter and said it would be better if I saw the badgers before they dispersed for the autumn, and so here I was. The kitchen smelt of cannelloni: she had cooked me supper and was going to put me up for the night. 'People think I'm mad,' she said. 'Until they see the badgers.'

If badgers were a problem on the farm, perhaps they could find peace closer to towns and cities. Here, at least, lived the badger-hugging majority who supported them. Away from the farm, we relate to badgers in the sentimental style of our time: we watch and admire them, we give them names, we bribe them to come closer with gifts of food.

Badgers are thriving in towns and cities such as Bristol, Brighton, Birmingham, Edinburgh, Exeter, Northampton, Nottingham, Swindon and Southend-on-Sea. Unlike urban foxes, these street-smart badgers have not taken up city life to cadge food from the allotments and patios. Although a few rural badgers may move into the suburbs, it is more likely that they have been there all along, surviving in the cracks between the highways and housing estates built around them. Urban badgers are mostly residues of historic rural populations. 'People talk about urban badgers but I call them urbanised badgers because they were here before human development and they've stayed and tried to continue,' said Don Hunford, an elderly resident of Benfleet, in suburban Essex, who feeds badgers in his back garden. 'They are still holding on to setts which are now completely urbanised. They don't like moving. They'll cling to ancestral ground.'

Suburbia is far from an uncontested space for badgers. Even in an environment under the totalitarian rule of humankind, some perceive any nuisance from a rival mammal to be a threat. Suburbanites become instinctive farmers as soon as their prize vegetables are dug up; plenty of allotment holders wish they could shoot the badgers that rob them of their runner beans. In Sheffield, residents appealed for something to be done about the 'sex-mad badgers' who howled and screamed while having sex and made 'terrible bloodcurdling noises' when they fought. In Evesham in 2007 it was reported that 'a rogue badger attacked five people during a 48-hour rampage in a quiet suburb'. Each year hundreds of setts are closed or relocated under government licence because they obstruct new developments or undermine roads, railways and even the foundations of houses. Despite these constant strikes against the badger, relic populations in towns have grown in recent decades. One reason may be people like Judy Salisbury: feeders.

*

When Judy moved to her house twenty-six years ago, neither she nor Robin had ever seen a badger. One evening, Robin was watching the news when Judy, sitting in front of the big window, spotted a badger crossing the lawn as if it owned the place. She wondered where it came from, and whether it would come again. The following dusk she saw the animal once more, and threw it some peanuts, which were devoured. For a few months, the badger vanished. 'When she came back she had two little heads with her. Twins,' said Judy. 'She brought these babies straight to the French windows to show me and it just went on from there. I started feeding them with a long-handled wooden spoon. Gradually I brought the spoon closer and closer. One of the babies was the first to take food from my hand. Dainty was the name of the mother. We called the cubs Rough and Tumble because that's exactly what they did.'

Judy took me to her bureau, wobbling slightly because she could not bend one knee, reached down stiffly, and pulled out a heavy photograph album. 'DAINTY,' it said on the cover in neat handwriting, 'To whom this album is dedicated'. The photographs, from the late 1980s, were fading but Judy looked exactly the same, her hair just a shade darker. The badgers had first been watched in October 1987 and two cubs arrived the following spring. The next year Dainty had four – Bertie, Gert, Daisy and Pipsqueak. Judy could easily tell them apart: she noticed their different characters, markings and colouring. Some were brown; some were grey. One cub, Flo-Jo, was blind and apparently orphaned. Judy could see there was something wrong with her but the other young badgers accepted her. 'They used to look after her – they moved around with her and pushed her towards the food. She lasted a long time.'

Judy and Robin opened up their home to their rapidly expanding badger family. The pictures showed it: French windows cast wide in winter; Judy passing a piece of bread from her mouth to a badger's mouth; Robin in his armchair, badgers lunging at his slippers. The badgers would tear around the living room playing, seizing shoes and taking them back to their sett.

Those who befriend wild animals must learn to cope without good-byes. Badgers are not companions. They do not lie down in their basket, lick their mistress's hand and go to sleep for ever. They only disappear. Grandma struggled to come to terms with the way Bodger vanished and, with characteristic pessimism, filled in the blanks by convincing herself it had been in agony. For Judy, Dainty's children just trotted off and never came back, mostly in the winter. At least Dainty plodded on, living to the grand old age of thirteen and bearing five sets of cubs. Then, she broke her back. 'She came here in distress dragging her hindquarters. There was nothing I could do. You can't take a wild badger to a vet. It's best left to nature,' remembered Judy. For more than a year, Dainty hobbled into Judy's garden every evening. The crippled badger struggled to climb the steep grass bank in the garden so Judy had a concrete ramp made especially so Dainty could climb up to the patio and receive her evening meal. 'I sat with her while she ate. One night she didn't come. She would've been buried in the sett by the others.'

Judy fell silent and then announced that she would change into her badger kit. Left sitting alone, I realised her house took me back to my grandparents': a certain smell, of furniture from the middle of the last century and a tinge of damp; a sense of things lying, well dusted but undisturbed; and a peace unbroken by those half-heard electrical hums from computers and chargers and Wi-Fi that fill more modern homes. Judy had one radio in the kitchen and a television that spent its days switched off. Outside, a herring gull called across the estuary and a wood pigeon thrashed about in the hedge.

When Judy returned, she was wearing a blue boiler suit – her badger-feeding uniform. When she offered one to me I felt obliged to accept. From a practical point of view, it might mask my alien smell from the badgers.

'Have you noticed how quickly the tide has come in?' asked Judy when I was changed and ready, squeezed into a blue nylon suit of the type I had not worn since I was a boy, mucking about on a farm. I had blinked, and missed it. The muddy estuary had filled up, as quickly as running a bath.

The banquet laid out, we sat and waited for badgers. We talked, desultorily, as the sea came up and the sun went down. Judy was not raised in Cornwall; she had been born in the Cotswolds, which seemed to be the unofficial capital of Badgerland. She was sent to boarding school when she was four years old. 'It was too young, really,' she said, with the understatement of an older generation. Her sister was six years her senior and they grew up not really knowing each other. Judy adored her father, who went into banking rather than take over the family farm. 'I didn't see very much of Daddy because of the

Second World War. He was running the head bank in Oxford and then he'd have to go on fire watch.'

Judy had always got on with wild animals. 'I suppose it was a gift I was born with. I was a very lonely child and I used to go into the woods and build little huts as children do. I got used to having wildlife around me – foxes and hares.' Despite playing in the woods of the Cotswolds, she never saw a badger.

Judy settled in Cornwall with her first husband, Peter, a military man. The story of how they met was very romantic. Following a serious illness, Judy travelled to Germany to recuperate with her older sister and her husband, who was head of the intelligence corps there. On her final evening, the commanding officer's wife threw a farewell party. Peter was detailed to pick up Judy and drive her to the party. 'He was fed up because he had a girlfriend but the CO's wife didn't like her and didn't invite her,' remembered Judy. Peter rather monosyllabically carried out his duties, ferrying Judy to the party. Afterwards, everyone went to the officers' mess and someone struck up on the piano. Judy had no partner; Peter dutifully asked her to dance. 'The moment he took me in his arms to dance shivers went through both of us and we did not stop dancing,' remembered Judy. 'It got to twelve o'clock. After we left the officers' mess party Peter and I went on to a nightclub. We both loved dancing. We danced until four in the morning.'

Judy caught her plane from Düsseldorf at 11 a.m. the next day. Before she left, Peter asked if he could visit her in England. Her sister gave her a look and said: 'You'll marry that man.' When Judy met

Peter at Oxford railway station, she got a shock. He was very tall and handsome, she conceded, but he looked silly in a pinstriped suit with a brolly, a briefcase and a lot of luggage. Worst of all, for Judy, he was wearing a bowler hat. She had always loved cars (Peter never cared for them) and was driving an open-top BSA, a sporty two-seater. 'I looked at him and said to myself, "I can't drive all through Oxford – people will see."' She was so embarrassed she zipped down the back-streets so no one would spot her with this bowler-hatted gentleman who had so obviously dressed to impress. Despite that embarrass-ment, they married in 1953, after he served in Korea. They retired to Cornwall and after Peter died, Judy met Robin, who had moved to the county to take over his family's holiday home. She never had children. 'I was very lucky,' she said. 'I had two very happy marriages.'

'Nearly time,' said Judy, glancing at the grandfather clock. It was five minutes to eight. A feral cat named Treacle arrived on the patio, expecting to be fed. I wondered if the cat would scare off the badgers. When Judy rose and offered the cat a dish containing a morsel or two from the badger's feast, Treacle held out a paw, as if it was saying thank you, or craving the touch of a human hand.

The cat disappeared again. A few minutes ticked by. It was still broad daylight outside. Suddenly, a badger. It thrust through the clotted-cream-coloured elder blooms and bounced up the lawn like a little trotting horse.

Judy slid the patio door ajar.

'Come along,' she quavered. 'Come along then. I can see you.'

The badger stopped, looking towards us, ears thrust forward like miniature satellite dishes.

'Here's one to take back,' said Judy, throwing a sandwich. 'You can find it.'

A second badger emerged. This one moved up the path towards the patio like a bow-legged old gardener before turning and bolting in a suddenly acquired sinuosity of movement that more resembled a rabbit. Fifteen seconds later, it bobbed out again, before skittishly dashing for cover once more. Phil Drabble once admired the badger's 'springy bear-like gait' – which was 'leisurely but deceptively efficient' – but these animals rapidly switched between lollops and gallops.

Judy hurled more sandwiches onto the patio with a child's green plastic spade. In a flash, I could see her as a young woman, headscarf flying, driving devil-may-care fast through the lanes in her sports car. There was nothing of the timorous elderly lady about the way she threw food to the badgers.

Like the incoming tide, unnoticed, four badgers had arrived on the patio. These were the descendants of Dainty. Today's matriarch, and the badger Judy knew best, was Willow. She was paler and browner than the others, with a broader stripe down her nose. This year's cubs were Salt, Pepper, Mustard and Vinegar and they looked like surreal garden ornaments, pausing, and raising their noses to the air.

'Would you like a sausage?' Judy called out in her special, high-pitched badger voice. 'Willow, have a sausage.'

Willow stubbornly stayed five yards away, eating Judy's special badger mix with a wet noise, somewhere between the slobber of a dog and the tidy consumption of a cat.

'Come along, Willow.'

Willow cautiously approached the French windows. Judy held out her hand. Suddenly, very decisively, the badger took the sausage from Judy's fingers with her teeth.

'She'll probably come back again,' said Judy. 'Would you like to give her one?'

I held out a wobbly third of a sausage, which flapped slightly in the breeze. When Willow returned to the top step, even my feeble nose could smell the rich musk of badger. She seemed to elongate her expressive nose, snapped her jaw very precisely around my proffered sausage and whipped it away before I could blink. I was still flinching as she retreated down the steps, trotting, head held high, with her prize.

As bats took to the darkening sky, nine badgers congregated on the patio, merging and separating and rippling like grey waves.

'They are enchanting, aren't they?' said Judy, giving her arm a well-practised swing and throwing handfuls of peanuts onto the grass. The family bond attracted her. They appeared to be uxorious creatures, pairing up and staying together. 'The way the mothers treat the cubs, the way they groom each other. They look like they are in love.' It was difficult not to compare them to our domesticated friends. Sometimes they were snuffling piglets; at other times they slunk off like a disgraced dog. When one grabbed a sandwich and sped off another galloped alongside it just as a puppy would, trying to snatch food from its playmate's mouth.

The nine bodies eddied and swam this way and that, as if they were connected, a toy train of animals. It did not seem a natural scene. Badgers may live in mixed-sex groups in Britain, but they tend to be solitary foragers unless there is a glut of food. Most farmers, of course, would hate this kind of husbandry but many animal lovers disapproved of feeding badgers as well. What kind of dependency was being created? What would happen to the sett when Judy was no longer here? Judy's banquet was one hugely influential type of human–badger relationship, however. Here was a deep bond between an isolated elderly lady and a family of badgers and, like all things in Badgerland, it unfolded beyond the trundle of conventional life. 'A lot of people say, "Aren't you lonely down there on your own?"' said Judy. 'I've never been lonely in my life. How can you be lonely down here?'

Judy Salisbury's nightly feasts probably sustained a larger population of badgers than would otherwise live on a small peninsula well populated with dogs and caravan sites. Feeding had helped badgers survive; so too had supreme obliviousness. My senses had been blunted by sixteen years living in cities, and my own low expectations of seeing wild things there. I never thought to seek out badgers in the town. At first, Maureen Davies was similarly blind. Then, three years ago, she popped outside her door one evening to put some bits in the bin and saw an animal on the wall. What the bloody hell is it? was her first thought. She was scared, because it was big. She rushed indoors and told her husband, Charlie. He said, Don't be silly, it's only a badger.

Maureen and Charlie lived in a stone terrace in one of the shabbier corners of Bristol. The M32 thundered past; so did trains heading for Temple Meads. In front of their house was a cemetery. Maureen might well never have seen a badger in the seven decades she had lived in this part of Bristol and yet they had always been around, venturing out of setts in the shrubbed edges of a park. Rather taken with the exotic-looking animal that had given her such a shock, she began leaving food on the cemetery wall opposite her front door.

Over the past three years, she had developed a routine that was delightfully convenient for the badgers. Bristol's badgers wander an average of 1.2 kilometres each night, a shorter distance than any other measured population (in Poland, badgers have been found to wander an average of 7 kilometres), probably because they can get all the food they need from their immediate vicinity. 'I just chuck everything in together,' said Maureen, showing me an enormous saucepan. 'Stale bread, veg, anything left on the plates. Apples, cucumbers, potatoes, carrots.' Friends from a health food shop provided thirty-six boxes of out-of-date oat biscuits (organic, naturally). The badgers were still ploughing through these, along with dog mix and peanuts, which Maureen bought specially. The health food shop friends also offered a bag of spicy peanuts, which the badgers devoured.

While dishing up on the wall, Maureen would see the badgers sitting in the cemetery below, which slipped down the leafy but noisy valley. 'They'll watch me doing it and then they are up on the wall,' she said. Routine spells safety and the badgers were wary of strange occurrences; the previous night someone parked on the street in their old van and the badgers didn't like that at all. 'If there are too many people or cars going past they don't come out,' explained Maureen. But generally, since she began feeding them, the badgers had got much braver. Especially the Big Un.

I visited Maureen on a cool October night with my friend Jez, who lived near by and was hoping to take some photographs of badgers. On her hob stood her big pan, filled with a lamb bone and various fruit and veg: a stew for the badgers. Charlie hailed from the Welsh Valleys and remembered badgers from his childhood. 'There were always stories about people going out badgering.' Badger hunters would wear shoes and gaiters and, to protect themselves from a vicious badger bite, would fill them with cockle shells. If gripped by a badger's jaw, the shells would crunch. 'Badgers won't eat bones. Once he touches a bone, he lets go,' said Charlie. I'd heard something similar in Ireland; there, apparently, hunters put a twig in their wellies. If a badger bit their leg it would quickly stop when it heard the bone-like crack of a twig inside the welly. 'Ah,' said Charlie, weighing up twigs versus cockle shells. 'We were posher down in Wales.'

Charlie met Maureen when he worked in Bristol as one of 'Dr Beeching's boys', tasked with dismantling some of the railways the infamous chairman of British Railways, Richard Beeching, earmarked for closure in the 1960s: there were 110 shunting tracks at Barrow Road 'and we cleaned them out in two years,' said Charlie. Later he made plastic ashtrays for prisons and bred chipmunks and exotic rabbits. There was good money in chipmunks. Another enterprise saw their back garden turned into an aviary for finches. The couple were now retired. Their living room was dominated by the television, family photos and, in the old fireplace, a collection of contented buddhas. 'She was told not to polish them,' said Charlie. 'That made her day.'

'I love them,' said Maureen. 'I'm a weird bugger.'

Maureen and Charlie were probably the least eccentric of all the badger lovers I had met so far and, unlike my grandma, did not relate to animals first and people second. It was too glib to assume that feeders saw their badgers as surrogate children. The photographs on Maureen and Charlie's walls were not of badgers but of their grandchildren and five great-grandchildren. 'My family means everything to me,' said Maureen. Then again, her badgers were substitute pets of a kind. The feeding had definitely developed momentum since the deaths of Chloe and King, Maureen and Charlie's beloved Staffies. 'We're getting no more dogs,' said Maureen. 'It's heartbreaking when they go.'

It looked as if the first frost of autumn might arrive that evening as Maureen descended her front steps and crossed the street with her pan of food. She placed the lamb bone on the cemetery wall, topped with iron railings, underneath a street lamp. Between the railings, she distributed dollops of dinner in a line for twenty yards. The M32 hummed in the background.

'You can hear their nails on the wall as they climb up,' said Maureen, shivering with anticipation. 'It's lovely when they come out. I get quite excited.' We settled down to watch, Maureen and I perched on the low wall by her front door, looking across the narrow street to the cemetery. Charlie brought out a folding chair for Jez to sit on, and stood behind us in the illuminated doorway.

After five minutes, a fox sprang onto the wall and disappeared again just as quickly. A plane turned in the sky noisily and a second fox jumped up to pick at Maureen's stew. According to Charlie, one of their regular foxes was so big he was 'like a German Shepherd'. A passing car cast its headlights across the fox's alert face. There was a colossal bang from an early firework. 'That'll see him go,' predicted Maureen. But the fox licked its chops and stalked with deliberation along the wall, poking its foxy features through the railings towards us, puzzled by the sight of four humans huddled on a lighted doorstep.

Maureen's record was five badgers and three foxes feeding together on the wall. In a land of suburban plenty, the animals shared their food. 'The funny thing is, they've never been scared of each other. They've come for food and met in the middle.' Maureen put her hands flat and brought them together.

I was amazed that health-and-safety busybodies hadn't outlawed Maureen's nightly canteen, especially since she featured on *Autumnwatch* on the telly. Council officials did erect a sign saying, 'Don't feed the birds', but Maureen carried on because she wasn't, although a few magpies gathered at dawn to snatch any leftovers. Occasionally neighbours joined her to admire the badgers. Only one local resident 'started a slanging match' and accused Maureen of feeding the rats. 'I've never seen a rat there, not once,' she said firmly.

Studies of badgers in the city found that 50 per cent of their diet was scavenged, and Bristol's badgers only really caused a stir when they pinched sweetcorn from the allotments.

'People lost their plots because of the badgers,' said Charlie with a straight face.

Just then, a ludicrously white nose appeared, testing the air and dabbing at the food on the wall.

'He's actually hanging onto the wall,' whispered Charlie. 'That's the Big Un.'

A scrabble, a scrape, and the badger hauled itself up the stone wall, which fell away more than a metre into the cemetery. One of the badger's less heralded physical attributes is its climbing ability. Badgers scale plum trees for fruit in late summer; they have also been spotted climbing five metres up trees, usually to find slugs but, in one case, to rob a bird feeder. Scientists once set a cattle trough on telescopic legs to see how high badgers could climb: one individual shinned 115cm up the steel legs, hooking a forepaw over the edge to haul itself into the trough. Keeping badgers away from cattle on farms was not as simple as it sounded.

Parading along the top of the cemetery wall, Big Un really was a big badger. In every badger-watching book, writers gave 'their' badgers names, and it made me cringe slightly. I am frequently guilty of sentimentality but naming these animals seemed to belittle them. It suggested that we, the writers, were getting rather more from this relationship than the badgers. The food was a shameless payment for friendship, a bribe for the badgers to cast aside their right and proper caution and come near us, their only, and deadly, foe. They could never be domesticated but we ascribed human behaviours to them and turned them cuddly. Now I was spending time with residents of Badgerland who had ongoing relationships with these animals, I saw it was not so simple. If you lived alongside them every day and if they responded to you, each according to their own, distinctive character, how could you not regard them as individuals, with names and souls? The relationship between these feeders and their subjects was as emotional as a farmer's with his cows.

Scratch scratch. A second, fluffier, badger clawed itself up onto the wall. These suburban badgers were not manky like many urban foxes. As a police helicopter hung in the distant sky, the fluffy one turned its broad head through the railings to face us. For an unnerving moment, it looked as if it had been separated from its body, stuffed and mounted on a plaque.

The pair steadily consumed the food while two men lingered at the corner of the cemetery. 'I hope that isn't the yobbos come to sell drugs,' said Maureen. A bike's flashing light came into view and Maureen predicted the badgers would bolt. The cyclist sped past, turning to look at us but not noticing the badgers to his left. Neither badger leapt to safety. A car revved up and passed within several feet. Big Un froze and, when the car had gone, resumed his methodical feeding. The clop of approaching heels rang out on the pavement and two women walked by. Once again, the badgers did not budge. Next to pass was the click-click of a dog on its evening perambulation along the pavement. The pet and its master passed within three yards of one badger on the wall. Only the dog turned sharply and tested the air; the badger paused, alert, but did not flee.

'As a rule, the Big Un, he never moves,' said Maureen.

'He fronts them out,' said Charlie.

The Bristol badgers had grown accustomed to passers-by. Maureen and Charlie believed the creatures could distinguish between familiar dogs, which didn't scare them off the wall, and strange ones, which did; they certainly recognised Maureen's voice. This did not make them tame, however. These badgers were as alert to alien sounds as any rural badger, but simply possessed a different set of skills. At one point, Jez pressed the shutter on his camera. Both badgers stopped and turned to inspect the source of this minuscule, unfamiliar noise. When I put my cup of tea down on the wall with a faint chink against the saucer, the badgers froze again, and evaluated the air through which this tiny noise had travelled. The Big Un decided he didn't like our presence one bit. As if in slow motion, he dropped very deliberately out of sight into the cemetery. Later, when I crept across the road to have a look at where the animals were going, one squatted, a pale-grey lump between pale-grey gravestones, looking up at me, a magical sprite in a city cemetery.

The most amazing thing, however, was not the badgers' ability to assess risk in their suburban environment but the oblivious people who passed them. A bloke smoking a cigarette walked within a yard of Big Un. If he'd trailed his left arm by his side he would've stroked the animal. When he stepped past them, both badgers jumped off the wall onto the scrunchy carpet of dead sycamore leaves in the cemetery.

Only when the man heard this unmissable thud and rustle did he turn his head for a split second as he walked on.

This moment was a perfect portrait of the blunted senses of modern man but it also showed our vulnerability. Our reflexes are catatonic, compared with the speed and grace of the most blundering of wild beasts around us, so estranged are we from the natural world.

Then again, we are utterly indifferent to badgers when they pose no threat. If we were at risk from a beating by Big Un, we would take notice. Another man walked past. He instantly clocked the shadowy figure of Jez, who was now sitting inside my car parked on the street, from where he was seeking some badger close-ups. Jez looked like he could have been on some kind of stakeout. But the man failed to spot the badgers perched on the wall right in front of him. We notice other humans because they are the only potential menace. I guess you can't blame us. We are only following our animal instincts, after all.

The London clay of the Thames Valley is too heavy, and liable to flood, for a safe sett. Travel east, however, and a wooded escarpment winds its way from Basildon to Benfleet and through Southend to Shoeburyness. Its fine, sandy soil is perfect for badgers.

A few weeks after my trip to Bristol, when I stumbled through Don Hunford's doorway at dusk, I felt like I had fallen down a hole and rolled into a badger's subterranean chamber. It was so dark inside his home that I could not see where the kitchen ended and the walls began. Don, a former science teacher who had been watching badgers in his back garden for decades and was said to know more about their behaviour than many scientists, was somewhere over on the other side of the table. I had been keen to meet him but he emanated doubt. I sensed I was being evaluated by someone who could see much better in crepuscular conditions than I could. I waited for him to turn on a light but he didn't. His kitchen smelt of bruised apples, and, slowly, my eyes began to discern objects in the gloom.

There were no surfaces visible on the worktops or anywhere else across the large room. Piles of indeterminate dusty objects teetered on trestle-tables like unfinished games of Jenga: wedges of conservation leaflets that had not been handed out, an aluminium Longworth trap for catching small mammals alive, bits of wire and biros. There were boxes full of torches, corks and, invariably, peanuts. I think there was a picture of a badger mounted high on the wall but even its black and white badge failed to show up in the gloom.

Don lived on a rough gravel road close to the Bread and Cheese pub in Benfleet. This was plotlands country. The abandonment of the land in the decades after the agricultural depression of the 1870s left great tracts of poorer soil uncultivated. Dilapidated farms fetched bargain prices. Many working Londoners, now granted a precious few days' holiday each year, aspired to build a weekend retreat on small parcels of land going cheap in Essex. In this era before planning controls, plots were divided and subdivided and ramshackle hand-built chalets rose up. Cockneys, Gypsies, democrats and dreamers settled in Benfleet, Crays Hill, Jaywick Sands and Canvey Island. A hundred years on, and building regulations, affluence, and the bog-standard products of the big construction firms had not yet smoothed these communities into banality. They were not conventionally attractive but they were cared for, extended and fancified, expressing the individuality of those who built them.

Don was born in Essex but later moved to the Cotswolds, where he taught maths and physics close to Ernest Neal's old school. Every resident of Badgerland seemed to have a connection with the Cotswolds. Sometime in the 1950s Don tried to take a picture of a badger at night, and a new obsession was born. Later, when work brought him back to Essex, he chose to live here because he knew the surrounding woods held badgers. After some years, his old house was cracked open by subsidence and I was stunned to learn that his decrepit-looking 'new' home, hidden behind great hummocks of elder, buddleia and bramble, was only twelve years old. It looked ancient, unlike Don, who appeared far younger than his years. Without doing the maths, I had initially assumed he was in his mid-sixties and was amazed when he revealed he was actually eighty-six. An old man who looked young, living in a young house that looked old.

The wood around the house was 'part of the plotlands which people thought would turn to gold dust in their hands', mumbled Don, who swallowed his words so it was hard to hear what he said. Thirty years ago, he bought six acres of overgrown orchard and oak woodland next to his house so he could watch the badgers. Like Judy Salisbury and Maureen Davies, Don fed them every night, but served up a more ascetic repast of bread and cheese, and peanuts of course. He did, however, stoop to distributing a few Rich Tea biscuits. 'I have a bit of a conscience about giving them biscuits. After all, they don't have dentists to go to,' he said. 'But they regard it as a bit of a reward at closing time.'

Don plunged into the wood that had swallowed his house and after a short tramp downhill we stood in an orange dusk. Security lights from the hefty mansion opposite Don's – a plotland chalet turned into a Roman-pillared villa fit for a footballer – filtered through the trees, giving the wood a suburban glow. There was a rustle of footsteps on fallen sycamore leaves behind us. I turned round, but it was nothing. I had a curious sensation that we were being watched, which I quickly dismissed. Another rustle and I twisted around again. An immaculate red fox, his big bushy tail poised expectantly behind, stood two paces from us. Just as I had become aware of the owl in the wood in Wales without really knowing it, here again I felt a dim flicker of some animal instinct that still existed inside me.

'That'll be Duty Fox,' said Don as if the appearance of a wild fox at our feet was the most natural thing in the world. 'She's looking for peanuts. She knows she'll get a handout.' Don's dusky universe was another parallel world. Ahead of us, in a clearing, stood a wooden box-shaped building, painted green. Sturdy, and built to Don's exact specifications, his badger hide featured a large wooden shutter that swung up to create an open window six feet by four, looking down onto the woodland floor. I reached inside the dark hide to help Don with the shutter's catch. 'Don't help me unless I ask for it,' he chided. I was reminded of Grandma: things must be done a certain way and other humans tended to do them the wrong way. It dawned on me that Don might have known Grandma. 'I remember Jane from the

Mammal Society,' he nodded. He once gave a presentation to the Mammal Society's annual conference called 'Myths about Badgers', but he volunteered no further information. I felt a twinge of desire for the gift of another memory of Grandma from Don, but something stopped me probing for more.

Within thirty seconds of our clumping into the hide there was a badger below the window. 'Looks like Jane,' murmured Don. Not my grandma – his badger. 'Oh, there's Elaine as well.' He need not have whispered, for these badgers were unafraid of his voice. They stumped closer. Elaine, a tiny creature who had been the runt of the litter but was now seven years old, placed her paws on the wooden walls of the hide, stretched upwards, and Don leaned out and casually fed her by hand. The badger did not snatch and retreat as I had seen at Judy Salisbury's, but took the piece of cheese very carefully before emitting a greedy glug-glug-glug as she wolfed it down. Don threw out small handfuls of peanuts. When they had been vacuumed up, a young boar called Oscar lay down and rested his nose on the earth, lying beneath the hide window in an expectant V-shape like a dog. 'They are just waiting around for a handout,' said Don. I had never seen such a relaxed group of badgers.

As more arrived, Don explained there had been a big fight among the badgers about two months ago. Since then he had not seen Hoppity, who was named for the way she dragged her hind legs following a probable encounter with a car. The fighting was unusual and 'put everyone on alert,' said Don. 'One of them, Honey, had the most horrendous-looking scars on her neck but they're healing up well. Candy is still uneasy with Honey around, although the last time she bunked off when Jake appeared. Jake certainly isn't aggressive but he *is* a robber. He will steal food if there's any to steal.'

It sounded like an episode of *The Only Way Is Essex*, except that Don did not simply follow the soap opera of the badgers' daily lives; through his years of observation he had acquired an acute understanding of their behaviour. He questioned many commonly held beliefs, including some theories of his hero Ernest Neal, such as the idea that only the alpha male and alpha female in each group of badgers breed. Don thought this assumption had been borrowed from other species. 'It certainly doesn't square with my observations. You either have a poor breeding season in which only one or two eligible sows breed or you have a good one. The record number of cubs I've had in one year is thirteen from one social group.' He had never noticed consistently domineering or 'alpha' badgers. Instead, he said, there was an easy-going, first-come, first-served principle at feeding times. In species where there is a strict hierarchy, there are recognised signs of submission – the cowering of wolves or dogs in a pack is a classic example, explained Don. 'Now I've never seen that in badgers. Some will move away because they don't want an argument, but that doesn't mean to say they will lie on their back and cringe like a dog will to a superior.' Don's badgers did not seek to curry favour by grooming the dominant male or female, either. As Judy Salisbury also noticed, each member of a badger group groomed every other animal.

It was fully dark now and eight badgers were just about visible in the clearing under the faint glow of the neighbour's security lights.

Each time Don briefly flashed his torch, the badgers would be caught in various poses, like mime artists. Elaine and Candy began a mutual grooming session. Then Candy went over to Dennis and plucked at tufts of his hair as delicately as a kiss. 'This is the social bonding, the mutual grooming. You see a lot of it,' explained Don softly. It was hard to reconcile these apparently sociable animals with the 'primitive sociality' that academics spoke of, although Don recognised that the badgers were only gathered together here because he fed them. Like suburbia itself, this assembly of badgers was a contrived wildness, constructed by humans. We may have changed their behaviour too.

Don got up abruptly and walked out of the hide with a staccato gait not unlike a badger's, so as to feed his animals more directly. Two jumped up, front paws on his knees, to take bread and cheese from his open hand. Silhouetted in the half-darkness, it was an appealing picture. I felt a stab of envy. I wanted them flocking around me like a pack of Dougals from *The Magic Roundabout* as well, but I could not compete with Don's years of commitment to one group of badgers. The animals would scatter in all directions if I, bearing my strange scent, stepped outside.

'That's it. That's it,' Don called out to his badgers, the contents of his tub frugally distributed among all eight over the last hour.

Why badgers? 'If I was in another country it might be leopards or tigers or bush babies. Part of the interest is because they are social animals, living in groups, which is analogous to the human way of life. Except we've made the groups too big now.' Don gave a wry grimace – he meant humans, not badgers. His demeanour towards his

badgers was like a veteran teacher, wearily familiar with the rolling production line of pupils but still possessing enough passion for the job to
find that some of his charges stuck in his memory. 'You do get
attached to some,' he admitted. 'I was very sorry when Honey got all
those wounds around the neck – she's been a lively, confident character. And little Elaine. All the years I've known her. She had a very
troubled childhood with watery eyes and abscesses on her face but she
got over it all.'

After accepting a rather reluctant offer of a cup of tea, I headed to
the door. It had been magical to see so much badger behaviour, and
yet I was not sure how welcome I was in Don's private world. So
when I stepped beyond the threshold of his tumbledown home his
question came as a surprise. 'Would you like to visit again?' he asked.
It hung in the air, vulnerably. 'Of course,' I said. 'In the spring. I'd
love to.'

10

Bait

It was a crisp, clear Sunday in midwinter when the screaming started on Paradise Farm. Strolling with a friend by the River Derwent in Yorkshire, close to the historic site of the battle of Stamford Bridge, Robert Fuller was looking for otters. Hearing the squealing, and a volley of barking a mile beyond, he quickened his step.

Dogs talk to each other, like the four-legged humans many owners believe they are, and there is a clear difference between the bark of a dog in a fight and the yap of a dog looking on. Fuller could hear both types of canine conversation and, as he drew closer, he detected a more chilling noise, an agitated chittering interspersed with screams and squeals, like a pig being tortured. Fuller, a well-known wildlife artist, had spent more than sixty nights watching badgers in the rolling countryside close to his home in the Wolds, and immediately recognised the sound of a badger fighting for its life.

Before he got to the hedge and dropped to the ground to peer into the field, he understood, with a sickening tightening in his stomach, what was going on. 'Each time I read through my police statement, my skin crawls,' he explained when we first met. 'I thought it was happening, and then to squeeze through the hedge and actually see it was a major shock.'

Fuller had stumbled upon a badger bait in broad daylight, on a Sunday, close to a public footpath, in the English countryside, in the twenty-first century.

Covered in blood, two scarily large dogs – 'like something out of *Harry Potter*,' Fuller remembered – were playing a violent game of tug-of-war with an airborne badger. The badger was still alive, emitting its terrified chittering, and attempting to snap back. Several terriers raced around, snatching bites. Chunks of fur flew. Most shockingly of all, a group of about eight men stood around, watching. One encouraged his dogs to attack the badger. Others stood back, holding shotguns, laughing and joking, excited by the blood lust they were witnessing.

Fuller may have been an artist but he was also a big, powerful Yorkshireman who had grown up on a farm. He had tackled hare coursers before 'and they are difficult lads to deal with,' as he put it. 'It's all a case of numbers. It's like wild animals really.' There was only Fuller and his mate, but having set out that morning to photograph otters on the river, the artist had a weapon at his disposal: his 400mm lens. He retrieved his camera from his rucksack and, after a whispered call to the police, wormed his way through the hedge and began taking photographs. He got a picture of a man in a green fleece shooting jacket and camouflaged cap urging the dogs on. He snapped others who were standing back, entertained, their dogs tethered on leads. For ten minutes he watched as the badger was repeatedly attacked and bitten, and the men made no move to drag their dogs away. Eventually the wild animal was so badly injured it could no longer fight back and the baiters lost interest. The man who had been goading the dogs pulled them off and a second man, wearing the flat cap and thick cords of a countryman, stepped in and shot the badger. The squealing finally stopped.

The fun over, the men chatted as they packed up their Land Rover with their gear, including their special locator collar, a device that tells hunters exactly where their terrier is when it goes under ground in pursuit of hole-dwelling animals. Then, one of the men spotted Fuller, and immediately covered his face with an outstretched arm. The other men began pointing at the artist.

For a moment, each party weighed up whether it was the hunter or the hunted. Fuller feared he was the prey and he and his friend headed off briskly along the river. But the badger-baiting gang knew they were in trouble too. They set about covering their tracks: the shot badger was thrown into a hedge and the hole they had dug was filled in with soil and pieces of turf neatly placed across it. At the bottom of the hole was slung the corpse of another baited badger, a heavily pregnant female whose intestines had been scattered across the rough pasture, alongside the tail of a third badger and four tiny, bright-pink foetuses. The baiters split up, some jumping into their two vehicles; others hurrying across the fields. The gang did not yet know that the police were quickly moving in.

*

I had assumed that badger baiting in the twenty-first century was a complete anachronism, another of those cruel country traditions which, if not regulated out of existence, had withered away. Here, surely, was one positive consequence of our alienation from nature. But digging and baiting had endured, despite the Badgers Act of 1973. In *The Darkness Is Light Enough*, badger watcher Chris Ferris wrote a terrifying account of her encounters with badger diggers in 1980s Kent. Around the same time, a friend who grew up in Pateley Bridge, North Yorkshire, remembers her Care Bear wallpaper being defaced by her politically radical older sister, who got into terrible trouble for scrawling on the walls: 'Sex Pistols' and 'Stop badger baiting'. Over in Ireland, the illicit Munster Badger Baiting Championship was traditionally held on the day of the All-Ireland Hurling final when the Garda were busy with crowd control in Dublin.

Whenever I met badger lovers, I asked them about digging and baiting. In the south and the west of England, people said neither practice any longer existed. A few experts – former police officers, animal charities that treated badger injuries – insisted digging and baiting still went on, mostly in the north. One RSPCA officer mentioned he kept his eye on a magazine for terrier and lurcher enthusiasts called *Earth Dog, Running Dog*. Here, I discovered the most demagogic modern-day apologist for badger digging. In the 1980s, when he was a judge at Crufts, David Harcombe, the editor of this modest magazine, wrote a book called *Badger Digging with Terriers*. As he recalled, 'crackpot' MPs tried to ban his book and he was 'pursued by the trendy pansies of the newspaper and television world' and eventually ended up in court. Found not guilty by magistrates in Gwent of inciting people to dig, injure or ill-treat badgers, Harcombe reiterated his defence of the sport in a later tome. Discovering his address and phone number on an old BNP membership list published by WikiLeaks, I telephoned and wrote to him in the hope of hearing an account of hunting the badger directly from his mouth. Sadly, he was having none of it. As I learned from one of his books, he considered journalists made 'the serpent or the scorpion appear as more desirable bedfellows'. As a nature-loving *Guardian* writer, I must have seemed a particularly poisonous reptile.

Luckily, over the years Harcombe has illuminated the world of the badger digger in his self-published writings. Terriers were his great passion and he bred, trained and sold these 'grand little warriors', using them to flush out foxes for the local hunt. Before it was illegal, he also dug for badgers, a sport founded upon an intimate relationship between man and dog. Terriers were not bred to be 'brain-washed into mincing proudly about rings'; nor did they merely catch rats or rabbits, argued Harcombe in *The World of the Working Terrier*. The ultimate mark of a proper working dog was whether it would confront a badger in its subterranean chambers. 'He must go to ground. He must stay to ground. He must work, alone, in the darkness of the earth, and see his task through until the men with the spades reach him,' wrote Harcombe. 'Really you know, the terrier had no right *ever* to work a badger. For a little bundle of courage weighing between twelve and twenty pounds to go creeping into the bowels of the earth to face a twenty-five to forty pound leviathan, unhurtable, unstoppable, unworkable, was just not on. And yet, the little worker did just that.'

Harcombe harked back to 'the badger digging boom years' when his sport was popular and accepted. Much of his writing echoed that of historic defenders of digging, such as Jocelyn Lucas, particularly in his insistence that the long odds against a terrier – 'one little dog in pursuit of its enemy in conditions which give that enemy every possible chance' – made it a reputable sport. Harcombe had his own clear digging code. He disapproved of sending many dogs after a badger – it should be one versus one, and no lurchers to tear the captured animal apart. He criticised betting on digs, believed diggers should observe a summer close season to give badgers respite, and claimed that, mostly, a dug badger should be released. If a badger was wounded and had to be killed, it should be shot.

Like many diggers of old, Harcombe certainly knew far more about 'billies' than me. He condemned the ignorance of a pro-badger group, charged with relocating a sett, who had dug up a lactating badger and concluded she had eaten her cubs; they were more likely to be alive, but hidden, he said: sows often abandoned cubs when under siege, spacing them around the sett to give them the best chance of survival. Harcombe also wrote elegiacally about how he cherished all life as he grew older. 'Not just human life, but the life of the animals about us. The terrier. The fox. The badger. All have value and a place in the scheme of things and we who are, or were, involved in the ending of that life must be aware of our responsibilities to end it as painlessly and untraumatically as possible. We owe that much to the animal. We owe that much to ourselves.' He hoped to become the oldest working terrierman in the country. And when he died he would like his ashes to be scattered over a favourite earth: 'Let my mourners be the generations of foxes and badgers that follow over the years,' he said, 'joyfully walking all over me and treading my ashes "to ground".'

Here, arguably, was a kindred spirit: an idiosyncratic loner who loved animals, long evenings outdoors and the smell of wet earth. Both David Harcombe and my grandma lived their lives tilted against the assumptions of the majority. Both would have more in common with each other than either would share with the badger-oblivious inhabitants of Britain. And yet, even if I accepted the premise that it was legitimate to kill animals for entertainment and cast Harcombe's defence of a rough country tradition in the most forgiving of lights, there were holes in his defence of digging. He skated over what actually happened when a badger was caught, and how routinely it would be killed, and didn't appear to believe that horrific injuries inflicted on his brave dogs constituted cruelty, either. He insisted he abided by a set of rules, and yet revealed there was never any consensus over how diggers should behave. Without clear rules, there was little self-restraint. Harcombe detested the 'unfair' Badgers Act but admitted that 'an irresistible case existed for the control of activities as they were then unfolding'.

Far too many '"Cave-man" characters came into the sport,' he wrote. These 'nutters' caused badger diggers to become 'the lepers of field sports'. Harcombe claimed he never intended to encourage 'the scum terriermen who soil our pond', yet he seemed fêted in those circles. His first badger-digging book had become highly collectable, fetching £100 online. An elderly man now, he was still churning out his monthly magazine with its rants against immigration and nostalgic accounts of historic badger digs. His magazine also hosted a thriving web forum where boastful men, and one or two women, shared pictures of their dogs and the wild animals they dug, chased and killed.

Nearly a year after Robert Fuller stumbled upon the badger bait at Paradise Farm, the alleged participants were summoned to the magistrates' court in Scarborough for their trial. As a vicious wind eddied around the seaside resort's thinly populated hotels, the men were accused of wilfully killing a badger, digging and interfering with a sett, hunting a wild animal with dogs and causing unnecessary suffering to an animal. This was a bait of men, cornered in Court 1, a district judge sitting high above them. Sergeant Paul Stephenson, the investigating officer, sat discreetly at the back of the court; at the side, a journalist from the local paper took meticulous shorthand notes; and on two rows of seats upholstered in royal blue were five people of late middle age, badger lovers who outnumbered the two women who were the only family members to support the men through their disgrace. The trial had been set down for a week. 'Some murder trials don't last that long,' exclaimed one police officer. 'This is murder,' murmured a court official.

Caught in traffic on his way to give evidence, Fuller pulled up alongside one of the defendants. The man leaned out of his 4×4 and gave the artist the middle finger. Behind the defiance, however, the accused must have been dismayed at the sight of Fuller. They knew their trial rested almost entirely on the evidence of this one perfect witness. Without Fuller's testimony and photographs, there would be no case. The seven men's best hope was that this farmer's son and family man would be too busy, or too frightened, to turn up.

But he had. Two of the group, Christopher Holmes, a twenty-eight-year-old from a smart new housing estate in York, and Malcolm 'Mally' Warner, also twenty-eight and a close neighbour, decided the game was up and pleaded guilty on the morning of the trial, knowing this admission could reduce their sentence by a third. They were dispatched to await sentencing while the remaining five men lined up in court, in their best dark suits and ties.

First in line was Alan Alexander. He was a short thirty-two-year-old, with close-cropped dark hair and glasses. Nicknamed 'Bok' by the others, he seemed a sociable guy, always having a laugh during breaks in the trial. He was the first man Fuller had seen, goading the dogs to rip the badger apart. Next to him was Will Anderson, who was twenty-six and big all over: tall, broad, thick curly hair and fat sideburns. Anderson lived in Pickering and ran his own business called Spirit of Adventure, customising off-road Land Rovers. He followed the proceedings assiduously, flicking through the case notes he kept in a shiny blue box file marked 'VAT receipts'.

The next man was the odd one out. James Doyle was a thirty-four-year-old with grey hair and possessed the doleful manner of a school-boy who fears he is in deep trouble. He lived further afield, in West Yorkshire, and claimed to know only one of the men. At the far end was Paul Tindall, a well-built former nightclub doorman of thirty-three with short light-brown hair, broad features and a relaxed, almost reassuring demeanour.

Only one man appeared completely in command of the situation he was now in. The eldest of the group, Richard Simpson was a countryman who had trained as a gamekeeper. He had a shaven head, strong black eyebrows and owned terriers called Red, Bramble and Gin, and a white lurcher bitch called Charlie. He loved dogs so much that the number plate on his silver Land Rover read 'DOG'. During the winter, he was employed as a beater on shoots, helping put up pheasants in return for pocket money and a free shoot at the end of the season. In summer, he and his partner, one of the women looking on, took their terriers around the country shows; she did the grooming, he did the walking. He met my gaze, and the gaze of Sgt Stephenson, and looked back keenly. He would not be anyone's prey. His partner was similarly forthright throughout the trial, muttering and nodding in support of her man.

She and Simpson lived in a 1950s semi on a council estate on the west side of York, close to their friend Tindall, and Alexander, who was Simpson's half-brother. It was here where the men gathered one Sunday morning in January when Simpson organised their day of 'rough shooting', with the help of Anderson, who claimed he had permission to shoot 'vermin' at Paradise Farm. If Simpson felt contempt for all this judicial hand wringing over two or three dead badgers, he was too self-controlled to reveal it. As it was being explained in court how Sgt Stephenson was on patrol that Sunday when his radio directed officers to Paradise Farm to investigate reports a badger had been shot, I suddenly realised what avid hunters would make of this. To them it would sound as absurd as police racing to probe reports of a pigeon struck by a car. Many hunting enthusiasts saw no reason why an animal as common as a badger should be so protected in law.

Half a mile from the farm, Sgt Stephenson stopped where two colleagues had pulled over Alexander's Land Rover Discovery, with Tindall inside. Alexander had bloodstains on his hands. One of his dogs, a bull lurcher that had ripped the badger apart, sported a fresh wound to the top of its snout (in *Earth Dog, Running Dog* magazine, such scars are called 'medals'). When Stephenson asked how the dog had received its injury, Alexander said: 'It's that fucker,' and pointed to a third dog in the back of the vehicle. 'It's a bad bastard and I'm going to get rid of it.' Yet when the dogs were put together, the 'bad' dog made no move to lash out. Stephenson found the locator collar in the Land Rover. It was wet and muddy. Alexander told him they used it for locating ferrets when they were flushing rabbits out of their burrows. The collar, observed Stephenson drily, was far too big for a ferret, even if strapped around its waist.

At the farm, the sergeant found Simpson and Anderson had also been detained, by other officers. The latter told him he had the landowner's permission and had phoned North Yorkshire police that morning to inform them they would be shooting rabbits on the land. This is common practice so that if concerned members of the public report gunshots the police already know what is going on. Anderson admitted shooting a badger. 'The badger ran out in front of the dogs and they ravaged it. We pulled the dogs off but it was in a bad way and I put it out of its misery,' he told Stephenson. 'I'm not a badger baiter.'

A few fields away, Fuller had doubled back and bumped into a fisherman he knew. Together they had seen three others, James Doyle, Chris Holmes and Mally Warner, lurking 'furtively' down by the river. The trio soon gave themselves up to the police.

Searching the scene, at first police found only the badger that had been shot dead, bloodied and torn apart and cast into the hedge. They returned the next day with police dogs, the RSPCA and a middle-aged woman called Jean Thorpe, who ran a rescue centre for injured wildlife. If Yorkshire people call a spade a spade, said a friend, Jean calls it a fucking shovel. An imposing grey-haired woman, Jean tirelessly tracked, recorded and pursued digging and baiting, even training police officers and appearing as an expert witness in court cases.

One dead badger had been found at Paradise Farm. 'We needed two,' explained Thorpe. 'One can be a mistake. Two cannot.' With Thorpe's expert eye, the police discovered the tail of another badger, the scattered foetuses and then, at the bottom of the filled-in hole, the bloodied carcass of the heavily pregnant badger.

*

Over the course of a week and a half, District Judge Kristina Harrison heard evidence from Fuller, as well as police officers and Jean Thorpe. The judge also listened intently to the men's defence. They stuck, tenaciously, to a story that excused the carnage. It sounded like the men believed themselves and, after a couple of days, I almost believed them too.

They described their Sunday shoot like this. After enjoying rabbiting together the previous weekend, they set out from Simpson's house and spent the morning walking through miscanthus, the exotic species of spiny grass that can grow nine feet high, seeking to flush out rabbits. They shot four pheasants in a wood – illegally, since you are not allowed to shoot pheasants on a Sunday – and then headed to Paradise Farm. Simpson and a couple of others fired at pigeons around the barns, failing to hit a single one. At some point, Alexander's two dogs accidentally got hold of a badger. He struggled to get his animals off and Anderson shot the creature to put it out of its misery. Doyle did not see any of this. Tindall had his dog on the lead. Simpson was a late arrival and admitted his white lurcher, Charlie, briefly also got hold of the badger, before he pulled her off. The blood on Charlie's mouth, however, Simpson ascribed to cuts from the spiny miscanthus; the blood on his jacket must have been a pheasant.

At times, it was almost comic how little they claimed to have seen of the badger being ripped apart. Doyle was too busy doing the crossword, sitting shivering in Simpson's Land Rover because he had forgotten to bring his coat that day. Then he was too busy trying, and failing, to phone his wife because he was worried about what was going on. Tindall was more concerned about the marshy ground under his feet and the water getting in his wellies to notice what was happening. Simpson was fed up because he just wanted to shoot some pigeons and exercise his dogs.

At the heart of the case, however, was a mystery: the dog that disappeared. Fuller had spotted a Bedlington terrier, a distinctive, aggressive breed with the coat of a poodle and a snout more akin to an American pit bull. By the time the police photographed the men with their thirteen dogs, this one had vanished. Either it ran off or was successfully hidden by the men. Or it belonged to someone else. Other witnesses, such as the fishermen, mentioned 'eight or nine' men on the scene. And so a story emerged of the mysterious extra man, a gatecrasher to the men's rough shooting party, whom none of them would name. He supposedly arrived separately, in a silver Daihatsu, and, they claimed, owned the Bedlington. Alexander and Anderson said they were scared of this man because he had 'a reputation'. Simpson said he was not afraid of him but claimed he did not know him and did not want to know him. Who was this menacing character? Did he exist? Sobia Ahmed, the young woman who was prosecuting, was dubious: 'It is simply a red herring that has been included by the defence at a late stage in proceedings to divert attention from what they have done,' she told the judge.

Clive Rees, a charming Welsh solicitor who specialised in 'the defence of wildlife and countryside crime', represented two of the men, Alexander and Anderson. 'They weren't badger baiting,' he told me bluntly outside the court. 'I've never seen anyone found guilty of badger baiting – of putting badgers in a pit with dogs against them.' When I commented that there seemed to be a resurgence in baiting, he said softly: 'There hasn't been a resurgence. When I was a boy every man with a terrier in the valleys used to go badger baiting at weekends.' On this occasion, however, he was unable to build much of a defence. He talked of the judge needing to 'look inside their minds' and assess their 'capacity for evil' because, he admitted, 'Nobody is going to say the deliberate killing of badgers is a good thing.' He fell back on the argument that hunting could not be done accidentally and the prosecution had failed to prove that his clients had deliberately killed the badgers.

Doyle, who had a different solicitor, stuck to a subtly different story and told it in a rather more repentant style than the others. 'That day turned into a nightmare for me,' he said. A man with a soft face who looked close to breaking down, Doyle was friends with Simpson from a few years back but now lived with his wife and children in West Yorkshire. He owned two dogs, slender whippety things, which were more family pets than anything else, he claimed. He loved nothing more than walking them with his wife and children and, once a month, he'd go rabbiting. 'Grandad used to take me ferreting so I'm used to rabbits and what happens,' he told the court.

While the rest of the men insisted they had parked their vehicles by the Paradise Farm buildings, Doyle, crucially, remembered that he had sat in Simpson's silver Land Rover somewhere along the line of the tyre tracks found in the field adjacent to the badger sett. The other men denied they parked here. If they had parked there, it suggested they had badgers, rather than pigeons, on their minds because they had pulled up right by a badger sett. Doyle claimed he stayed in the Land Rover because he was cold until he heard so much barking that he stepped out for a fag. 'I also heard another noise. Something I've never heard before,' he said. 'All I can describe is it were like a small dog or a small animal in pain, screaming.'

Worried, Doyle said, he tried to ring his wife. 'I just phoned her for a bit of comfort, to be honest; the first person I think to ring is my wife.' He could not get any reception on his phone and so hovered around. 'I could tell there was something not right, and I didn't want to look. I didn't know where exactly the noises were coming from and if anything was getting hurt, a dog, whatever, an animal, I wouldn't be able to cope with that. I can't handle that. I wouldn't want to look,' he claimed. He then played some music on his phone, putting it to his ear to blot out the noise, oddly squeamish for someone who regularly watched his dogs kill rabbits. Summing up, Doyle's solicitor neatly described his defence. 'He can't become a party [to a badger bait], forgive me, Mr Doyle,' she said, glancing at her client, 'by being a coward.'

Simpson did not roll over in the style of Doyle and nor did he dwell on the cruelty of the death of the badger as Anderson did, speaking of the 'poor thing' being 'in pain' before he shot it. Simpson was unapologetic about his lifelong passion for country sports. 'A friend of mine's father used to go pigeon shooting, ferreting and whatnot and I tagged along and from that day on I've enjoyed it,' he told the court. He trained to be a gamekeeper at Askham Bryan College and, on the YTS, took his first job on a small estate near Pateley Bridge. He saw no need to pretend to be a bunny hugger – their day out was 'a rough shoot – pigeons, crows, foxes, stoats, weasels, anything vermin,' he said.

Ahmed, the prosecutor, asked why Simpson took Charlie, his lurcher, along.

'She was the best dog to take with me, that's why I took her,' he said.

And why take the gun?

'She can't catch pigeons,' he quickly replied. His partner grinned in the seat behind him.

Ahmed asked if Charlie was ripping at the badger with the other dogs. 'Not so much ripping, she just had hold of it,' replied Simpson. He clenched his jaw and a vein bulged in his forehead.

Was she ragging the badger around? 'No,' he said.

Were the other dogs ragging the badger? 'Yes,' he replied.

Did he realise it was a badger? 'Yes,' he said. 'I was present in the field but I wasn't watching the bait like you say I was.'

'The blood found on your jacket, that's badger blood, isn't it?' said Ahmed, with the practised accusation of a prosecutor.

'I don't know. I thought it was pheasant blood,' said Simpson before turning the questioning on his interrogator with a hint of enjoyment. 'Was there no DNA swabs taken then?'

The men must have known there had been no forensics undertaken. It was too expensive. Besides, the police thought they had enough evidence: the men had been caught, literally, red-handed.

*

A week later District Judge Harrison delivered her verdict. Doyle was found not guilty but the others were found guilty of every charge they faced: wilfully killing a badger, hunting a mammal with dogs, digging for badgers and interfering with a badger sett. A few weeks later, they were sentenced: sixteen weeks in prison for each of the four men.

It can be surprisingly difficult to obtain convictions over badger baiting and so the police, the RSPCA and Jean Thorpe were all relieved at this unequivocal verdict. The 'accidental' defence offered by the canny Clive Rees was a common one, and often successful.

Robert Fuller, whose photographs had made the prosecution possible, was glad to see that justice, for once, was done. 'I've got a shotgun certificate, I'm used to country life, I've seen a lot of things. When I was a boy, badgers around here were systematically gassed and there was no shame or embarrassment about it,' he remembered. The incident on Paradise Farm was different, however, and 'a horrendous thing to see,' he said. 'They were laughing as badgers were having their insides torn out by these dogs – and badgers are as tough as old boots, our toughest wild animal left. I cannot abide these people. You've got to stand up to it. It's unbelievably cruel.'

The bait at Paradise Farm was not an anomaly. I wanted to find out more about the baiting and digging subculture that persisted despite the law, and so I joined an Internet forum where terriermen discussed their love of hunting. In the past, baiters would meet in country pubs, their fighting dogs and baiting kit traded in newspaper small ads.

What was going on in social media was nothing new, except the authorities believed it encouraged people to travel further to torture badgers.

It also increased the potential for public boasting. Ian Briggs, chief inspector of the RSPCA's special operations unit and a man who has brought many baiters to court, believes digging and baiting are not primarily motivated by a desire to catch badgers but by an urge to show off their dogs. 'These guys are into their terriers or lurchers, they try to promote their dog as being "game" and they want to pit it against what is seen to be Britain's toughest mammal,' he said. 'It's bragging rights – posting trophy photos of themselves with a dog and a gun and ten hares or twenty rabbits.'

The online chat was typical of Internet forums: argumentative, cliquey, aggressive and dominated by a handful of regulars. Behind a keyboard, everyone is a hero. There was plenty of manly appreciation of those who posted photos of themselves in wintry fields, holding aloft dead foxes that flopped like rag dolls, their dogs standing alert beside them. Visible injuries were highly prized. 'That dog looks well worked,' wrote one man admiringly of another's photo of his heavily scarred terrier.

The Hunting Act that outlawed fox hunting also restricted the hunting of wild animals with dogs generally. It is still legal to send dogs after rabbits and rats, but nothing else. One dog can be used to flush a fox from its earth to a gun, but the fox must be shot and not ripped to pieces by dogs. The forum users were well versed in these rather complicated laws and seldom boasted of anything overtly illegal.

Although plenty hinted they might have tackled foxes with more than one dog, they were more careful when discussing badgers, which were euphemistically referred to as 'pigs'. The only badger-baiting pictures I found were posted by a South Korean, who after a few jokes about eating dogs was greeted with respect for extracting these animals from such stony ground.

Most British hunters were wise enough not to upload badger-digging pictures to public forums. But according to Ian Hutchison, a former police officer who is the UK crime prevention leader for Operation Meles, an anti-digging partnership involving every UK police force and leading charities, the use of mobile phones to film digs and the number of images found on the home computers of suspects is 'staggering'. This is animal torture porn. A few years ago in Northern Ireland, the Ulster Society for the Protection of Animals together with the *Sunday Times* set up a fake blog celebrating badger baiting, which ensnared a number of baiters. More often, however, the discovery of cruel pictures is accidental. When Northumberland police seized Wayne Lumsden's mobile phone in an unrelated investigation, they found video clips of dogs fighting other dogs, cats, foxes and badgers. The twenty-three-year-old from Lynemouth also posted footage on social media. Lumsden, and Connor Patterson, also twenty-three and from Northumberland, who had exchanged text messages about his enjoyment of a badger bait, were sentenced to twenty-one and sixteen weeks' imprisonment, respectively, in 2011. A month earlier, Christian Latcham of Tonypandy in South Wales was sentenced to five months' imprisonment, suspended for twelve months, after photographs and videos of badger baiting were found – again, by chance – on his mobile phone.

Today's badger hunter may market himself and his dogs on the Internet, but I was struck by how little had changed from the village baits described by John Clare. The main difference was that the modern baiter or digger usually came from a town or city, usually in northern England. Bok, Tiffer and Mally, as the baiters at Paradise Farm called each other, seemed to be good friends. All but one of the guilty men came from York, and most lived on council estates to the west of the city. 'I know it's not politically correct but it can be a housing-estate thing,' said Jean Thorpe. 'Country people do it quietly. You don't catch them. The housing-estate people tend to be gobby and shout about it and that's how you find out.' According to Ian Briggs of the RSPCA, the vast majority come from inner-city areas. 'It's very much a working-class pastime. A lot of these lads are involved in low-level crime,' he said. Often, claimed Briggs, such men would cast their eye over farm buildings to see what they might steal.

The badger protectors did not see any links between badger diggers and the scarlet-jacketed huntspeople who still ride out during the traditional fox-hunting season. Unlike fox hunting, digging and baiting have no place in any part of mainstream society, not even tucked away in an unapologetic corner of the British establishment. Unlike, say, the persecution of birds of prey, it is not condoned by a few big landowners. David Harcombe's out-of-print books are virtually the only ones explaining how it is done. It could now be celebrated in the wilder corners of the Internet, but the only way digging and baiting could have survived all these years is by being handed down, by father, uncle, cousin or brother, through the generations.

Ian Briggs of the RSPCA was in no doubt that baiting is a traditional sport passed through families of working-class men. 'They've been brought up around working dogs and they pick it up from their parents. They don't perceive what they are doing as being cruel at all. The fact that badgers have this protection, it's all these fluffy-bunny tree huggers stopping them doing what they want to do. They don't have dogs as pets, they have working dogs. Their view is that their dogs enjoy doing what they do. And a terrier does – it enjoys "going to ground" for its master. They see it as a lot of fuss over nothing – the badger is vermin, it spreads bovine TB. We've had photos of four- and five-year-old kids stood with dead badgers. Their Sunday morning is to go out and dig a badger and work the dogs.' Jim Ashley, a former policeman with an encyclopaedic knowledge of badgers in Shropshire, discovered a case where baiters dug out a sett he did not know existed. When he asked one of the men how he knew of it, the baiter told him he got it from 'my dad's maps'. The baiters had a better knowledge of badger setts than the conservationists.

Badger baiting is an activity that has been derided by the ruling classes and condemned by great poets and writers for 200 years. It has been illegal to bait wild badgers since 1835. In the last forty years, new laws have been passed in a concerted effort to banish baiting, and digging, for ever. These laws reflect the views of the vast majority, who believe they are barbaric activities. But still they endure. One of the solicitors who defended the Paradise Farm baiters shrugged and said this type of case would always occur because man, ultimately, is a hunter – and futile are the modern laws that try to suppress this nature. There are clearly intrinsic pleasures in baiting for some people: a day of tough physical work in the countryside; the thrill of doing something illicit; the excitement of your terrier, and watching her do the job she was trained to do; and the adrenalin triggered by cornering a truly unpredictable wild animal that could easily kill your dog or break your arm. But attributing it to nature being red in tooth and claw explains everything and nothing.

Badgers are still hunted in most other European countries, where they are treated as an ordinary game animal. In places such as Sweden, where badgers are plentiful and not much favoured, they are shot like deer and wild boar. However, the technique of digging badgers with terriers and then baiting them with other dogs appears to be particularly favoured in Britain (though in Ireland and parts of France too). Perhaps it has endured here in a unique form because of our class system. Poaching is romanticised because it was – and still is – a small way in which the rural poor got one over the men who owned the land. Badger baiting is similar. Men who have very little power in their working lives seize control in another sphere. It may be unspeakably cruel, it may be an expression of our basest instincts – man as bully, coward and thrill-seeker – but it is an expression of autonomy and freedom, and of one class's contempt for the laws made by another.

11

Patients

Behind automatic security gates in leafy Surrey, where dense shrubbery muffled the thwack of balls on the golf course, was a very expensive wire enclosure. It looked built to withstand a war and was equipped with central heating, lighting, an earth run decorated with tree roots and CCTV to carefully monitor the guests, who would be gently restored to a peak of physical fitness before being returned to the countryside.

This badger health spa stood in the grounds of a very beautiful and discreet Arts and Crafts-era house with rooms panelled in warm oak. On a perfect spring day, great tits see-sawed in the trees and Adele's latest ballad echoed tinnily from a builder's stereo as two men laid a stone pathway across a lawn the texture of a bowling green. In the old stables was an office where gold discs hung on the wall. After a few minutes, Brian May, the Queen guitarist, wandered in and greeted his assistant with a kiss on both cheeks.

I had only realised Brian was Britain's leading campaigner for badgers a few weeks earlier when a press release pinged into my inbox containing a soft-focus picture of the rock legend cuddling a badger. The accompanying words, unusually heartfelt for a piece of PR, alerted journalists to the fact that Brian was travelling to the Welsh Assembly to appeal personally to Assembly members to halt a proposed cull of badgers in Wales.

From his famous thatch to his immaculate white Puma trainers, Brian looked the quintessential semi-retired rock star. His black trousers were tight and his white shirt was unbuttoned to reveal a delicate gold chain and wisps of greying chest hair. Fast cars, metal detecting, saving the rainforest: ageing musicians enjoy indulging their passions but are derided for putting their names to causes they have the scantest knowledge about. Oliver Edwards, the farmer I met in Exmoor, was typical in his attitude towards the Queen legend. 'Stick to your guitar, Brian,' Oliver said when Brian's campaigning cropped up in our discussion of bovine TB and badgers. 'I'm not interfering in your way of life. I'm not interfering with your business. You're good at playing your guitar; I'm very good at farming. So let me carry on with what I'm doing as a conservationist and a farmer.'

Badgers and Brian May might seem one of those comically incongruous marriages of convenience between showbiz and a worthy cause. But it quickly became apparent that this was not the dabbling of a dilettante.

'Hobby, hmmm,' murmured Brian, toying with the word with a quiet contempt. 'Well, there aren't many people from rock music or entertainment who put the time in that I have. They'll put their name to something but this has become a huge part of my life. I don't care what people say. I know why I'm doing it. I'm not doing it to make money. I'm not doing it because I want to be famous. I'm doing it because I care about animals. This is not just something that concerns farmers; it concerns us all.'

As I had seen, badger feeders often treated their animals like members of the family. Another group of people went further, and sought to rescue and rehabilitate wild badgers. My grandma did it on a modest scale but these days there were badger saviours who were much more ambitious. Brian May, with his rock star millions, was one. Pauline Kidner, a farmer's wife who set up an animal hospital called Secret World at her farm in Somerset, was another. If your hours are devoted to rescuing individual animals, it is probably a logical step to seek to save a species, and just as my grandma's sleepless nights caring for Bodger seemed not exhausting but galvanising in terms of political activism, so Pauline and Brian, and the charities they had created, became influential players in the growing controversy over the badger cull.

When I first met him at his home, Brian May was still better known as the guitarist behind eighteen Number One albums and ubiquitous anthems from 'Bohemian Rhapsody' to 'We Will Rock You' than as the face of the anti-cull campaign. But Brian had always been a polymath with a broad hinterland beyond the world of rock. He abandoned a PhD examining zodiacal light to commit to Queen and later became a great friend of Sir Patrick Moore, the astronomer; together they wrote a history of the universe. In recent years, Brian returned to his abandoned PhD and completed it, and pursued a passion for 3D, or stereoscopy.

His flamboyant appearance led me to expect an opinionated showman, but he spoke softly and crooned 'mmmmm' when he agreed with something and, confusingly, made the same murmur with a subtly different cadence when he did not. He simultaneously combined the otherworldiness of a long-term resident of Planet Rock – an Ozzy Osbourne, say – with the sharpness of a professional scientist. He was also acutely sensitive to what he pessimistically saw as the brutality of the world. 'There are a lot of nasty people out there,' he said. He had met them on the Internet. Some time before it was fashionable, he created what he called his 'soapbox' – a blog. 'It changed my life completely because it's a two-way communication,' he said. The online conversations he began having with fans awoke his childhood passion for animals.

But Brian's animal activism also sprang from a profound sense that *Homo sapiens* is not the centre of the universe, a conviction derived from his astronomy. As he put it in another interview: 'For a thousand years the Ptolemaic system was believed all round the world. The system says that the Earth is at the centre of the universe. It turns out not to be true. But this idea that we are the centre of creation lingers on – where is the justification to say we are the central piece of evolution? There is nothing to tell us that whatsoever. So why would we use that to justify our exceedingly bad behaviour?'

Brian always promised himself that if he ever had an opportunity 'to make a difference for animals', he would take it. That chance arose when he was alerted to a proposed cull of hedgehogs on the Scottish island of Uist. He successfully fought to have the hedgehogs, blamed for the loss of birds on the island, transported to the mainland. A bridge was crossed, as he put it, and he was now in touch with just about every animal charity there was, from saving whales to 'Virginia McKenna fighting for elephants in Africa'. Well before the badger cull was proposed, he set up his own charity, Save Me, dedicated to fighting for all wild animals in Britain. 'It's been a journey and you start to question so many things, like the word "vermin". What does that mean? Does it mean an animal that is a nuisance to you? Yes. Does it mean that the animal has no right to be treated decently? Well, no. So many people have said to me, "Foxes are just like rats, who gives a shit?" So a rat isn't worthy of some consideration? Rats are so like human beings, it's frightening. Relegating them to vermin gives people the feeling that they don't have to treat animals with any kind of consideration whatsoever, and to me that's wrong.'

Without really seeking to, Brian May had become a spokesperson for badgers. In 2010 he joined the ultimately successful campaign against the badger cull in Wales; two years later, he led an alliance of animal rights groups opposing the cull in England. In doing so, he represented a view of animals held by thousands of animal rights activists and millions of other animal lovers who were not active campaigners.

The philosophy of rights for animals, an idea first taken up in earnest in the eighteenth century and given physical form with the creation of the Society for the Prevention of Cruelty to Animals in London in 1824, was not something I had previously given much thought to. I was passionate about our countryside, and the species that belonged there, but I tended not to consider animals as individuals or souls. My attitude was shared by most ecologists and conservationists, who treated species differently and gave certain rare or native species special protection. Animal rights activists thought that apparently privileging some species above others was illogical and discriminatory, instead arguing that every single individual animal had moral and legal rights.

'To me it's a basic assumption: animals have as much right to live on this planet as human animals,' said Brian. 'A decent life and a decent death – that's what I ask for myself, and that's what I would ask for any creature.' He argued he was not fighting for badgers on sentimental grounds, although his stance was certainly an emotional one. 'I don't really love badgers because they are furry and good-looking. They *are* appealing, there's no doubt, they're like little bears, especially when they're young. They are fascinating and rather mysterious because they've been around in the British Isles longer than humans and they have their own social ways, not all of which are understood by us. I can't help but have an awe for all wild creatures who have survived the awfulness of what we've done to the world.'

When Brian disappeared to take a phone call, I stepped into his garden to meet Anne Brummer. She ran Brian's animal rescue centre, the high-tech badger sanctuary where sick and injured badgers found in the wild were taken for recuperation. After they had built up their strength in captivity, they were returned to the Surrey countryside through a careful process of gradually allowing them to wander further from their pen. So far, no young badgers had been handed in this spring and Anne was instead caring for five fox cubs with soulful grey eyes and ears so soft they could not yet prick up. Brian had named the cubs after 'animal lovers' – coalition politicians including David Cameron, Nick Clegg and Caroline Spelman, the Environment Secretary who announced in 2011 that a badger cull would proceed.

The shrubbery rustled and shook as Brian rejoined us, stumbling out of the rhododendrons, the branches tangling in his grey curls. 'It's disappearing now. The hair has seen better days,' he said in an Eeyorish aside. He smoothed himself down and permitted himself a cuddle with the cubs at play. My grandma's philosophy of not allowing wild animals to be tamed was now the only acceptable stance in rescue centres across the land, and Brian usually kept his distance from the animals in his care. But he believed his charges made their own decision to break free of human bonds. 'You have to bottle-feed these little things,' he said, picking up Caroline Spelman. 'They seem to love it, but there comes a point when, just like children, they don't want to be with mummy any more, so they seem to naturally draw away. It's good for them. We don't want them trusting humans because,' he said, speaking to the cub, 'not all humans are trustworthy, no-o-o-o.'

*

Brian's campaign against the badger culls in Wales and England attracted opprobrium from farmers, and then rage when he suggested that farmers in bovine TB hotspots should simply stop rearing cattle, and cultivate something else instead – or move. 'Why is the taxpayer paying for farmers to bring cows into the world in TB hotspots?' May asked. He was 'severely attacked' for this idea and seemed visibly hurt by the criticism. 'There was a nasty little piece which said, "It's like telling Brian May he can't play guitar" and they said, "Some people would welcome that." Ha ha, lovely. But there was a time when Queen was very uncool in Britain and what we did was play elsewhere and I actually took my family to America and my little boy went to school in LA partly because of that. So [farmers relocating is] not such a ridiculous suggestion.'

The hostility to Brian reflected what was portrayed as a cultural tension between the country and the city, even if a majority of country dwellers were almost certainly on the side of the wild animal, as with the debate over fox hunting. Critics characterised Brian's pronouncements on badgers as just another showbiz townie from Millionaires' Row telling farmers what to do.

He grew up in suburban West London and his current home in Surrey was not considered proper countryside by people with mud on their wellies. However, his defence of his position as an urban-based critic of badger culling was part of a long tradition of intellectuals arguing that country people do not have a monopoly on understanding rural life. The first people to speak out against animal cruelty in eighteenth-century Britain were clergy, politicians and urban thinkers.

'Often the people most indifferent to nature and the most ignorant about all living creation, are those who dwell closest to it,' wrote Norah Burke in the 1960s. 'One can think of urban naturalists who know far more than many country people.' Brian made a similar point. 'If you want to understand the countryside, you have to come from the town,' he said, quoting an old adage. 'Tradition can be a terrible evil. People think, "Our family comes from here, we've always done it this way so it must be right." If that was true then we'd still be in the midst of slavery, we'd still be burning witches and we'd still be bear baiting. Things have to be re-evaluated and very often they have to be re-evaluated by people who come in from the outside and say, "Hey, that doesn't make sense. Even though you've been doing it for all these years, it's crap."'

Far from being out of touch, Brian reckoned that his blog put him in contact with both sides of the debate, including farmers. 'This old accusation that I'm not living in the same world – it's not true any more,' he said.

Brian had a strong grasp of the science of culling and was determined to push for a scientific fix for bovine TB – a vaccine for both cattle and badgers. But his beliefs about animals were very different from the dispassionate view of most biologists and ecologists who think of ecosystems and species rather than of individuals. Like other animal activists, he saw the cull as part of an irrepressible strand of cruelty in rural life, and, politically, he agreed that the cull was the Conservatives' sop to country voters because, locked in coalition, they lacked the numbers to repeal Labour's hunting ban. 'It's a panacea that is being offered to farmers – look, we are doing something, we are on your side, we're going out and killing things,' he said. But he also made a connection between culling and the digging and baiting of badgers, which he had seen evidence of online. 'There is a link. You go on these websites and the same kind of gung-ho "let's kill, kill, kill" comments come from people who are supporting the badger cull as from those supporting the repeal of the Hunting Act. I find that shocking. There are people who really take pleasure in killing.'

Brian did not appear to get much sleep at night and I wondered if he spent too many evenings confronting virtual savagery on the Internet, which does not always spill over into real life. 'It's very uncomfortable to see into the minds of people who are so full of violence and the need to cause pain. It's very upsetting and you have to fight being depressed about it. Some mornings I get up, I find it hard to deal with.' He paused. 'I get over it,' he said more lightly, and laughed at himself for the first time.

During our chat-and-play with the fox cubs, Brian fell silent for what seemed like an age when I asked him about the ethics of other culls, and if it was important to cull grey squirrels in Britain to give the indigenous reds a chance to survive. 'You can look at numbers or ecology, a rather flawed science in my opinion, and you can look at aggregates. If you're not taking care of the welfare of animals as well as numbers, you're not doing a good job. Every animal is an individual and if a grey squirrel is born in this country it's not its fault. It's a very dangerous premise to say things ought to be restored to the way they were a hundred years ago, because if we did that we'd be kicking all the black people out of Brixton. It's a fascistic attitude.'

I am no scientist, but I felt more inclined to value balanced eco-systems. Grey squirrels, introduced into Britain by man and out-competing native red squirrels as well as spreading a disease that was fatal to them, were not human immigrants; this was an anthropo-morphic comparison. Then again, perhaps our ecosystems were so unbalanced now anyway that it was futile to attempt to rebalance them by culling non-native species; for sure, our attempts to restore a 'natural' balance or control disease in the countryside had been pretty clumsy in the past.

The philosophy of animal rights held by Brian and many others was certainly coherent and it was also rather black and white. I wondered if it was right to apply it to every messy situation where the interests of different species collided. It was hard to disagree, however, with the words that lingered in my mind long after the automatic gates slid shut on Brian's secluded kingdom. 'We're fighting the symptoms – fox hunting or badger culling – but we're also trying to fight what's under-neath it,' he declared, 'which is this mentality that says human beings are the only creatures on this planet who matter.'

If Brian May's badger sanctuary was a luxurious Harley Street clinic, then Secret World, the biggest badger-rescue centre in the country, was a busy A&E. Every year fifty cubs and seventy adults from across the land were whisked to a graceful and slightly dog-eared seventeenth-century farmhouse on the Somerset Levels, where they were expertly cared for by Pauline Kidner and her staff before being returned to the wild.

As I waited for Pauline in Secret World's gift shop, one of an apparently limitless set of outbuildings mostly converted into wards, surgeries and operating theatres, I imagined she would be a determined zealot who had sacrificed everything for animals, but when she arrived I liked her immediately. Pauline possessed that no-nonsense approach shared by those who work with sick creatures every day, but she did not appear disillusioned with the human race. She was also full of good sense. The urban myth that roadkill badgers had been secretly massacred by farmers was 'absolute rubbish', she said: farmers had so much land on which to dump a carcass they didn't need to put it on a main road. Pauline had no time for badger feeders, either; she believed that people who befriended wild animals did more harm than good. Besides, she was once bitten by a badger, an excruciatingly painful event, and she knew they were not to be cuddled.

Back in the early 1980s when she was a young single mother, Pauline answered an advert for a housekeeper for Derek, a dairy farmer, who was also single and had children of his own. She moved in, and they fell in love. After they married, Pauline helped oversee their transition from a dairy farm to an open farm for holidaymakers, and then a wildlife rescue centre. She took me through a maze of passageways into her kitchen, which smelt of dog and was adorned with every conceivable kind of badger tat: badger plates, a badger clock, ornamental badgers on the mantelpiece. They were gifts and thank-yous from people she had met and she didn't have the heart to throw them away. As we chatted over tea, Derek, a tall, kindly-looking man, arrived to put something down for the mice that overran the kitchen. Tolerance of animals has its limits.

Pauline took in their first casualty in 1986. 'It was birds. Starlings,' she said. The local council decided to reroof its houses just as the starlings nested, and residents were surrounded by helpless chicks. Pauline and Derek still had incubators on the farm because they once kept poultry, and so they took in the nestless birds and saved their lives. As my grandma discovered, word quickly spreads if you take in rescued wild animals, and Pauline was soon inundated. And just as Grandma had been, her passion for saving injured animals was properly ignited by the charisma of the badger. Pauline's first, Bluebell, arrived for treatment in 1989. 'People love the otter cubs but I still prefer the badgers,' she said.

Secret World was far from a lavish operation but running an animal hospital is not cheap: Pauline's costs were £500,000 each year. Unlike Brian May, she had no personal fortune to draw on and so all of this was raised through individual donations. The fact that she was still going strong in the midst of a long recession demonstrated the enduring attraction of animal charities for the British public.

Pauline's individual donors supported an operation that did not discriminate between species and sought to save every animal admitted. Each one was recorded in a logbook outside the old cheese room. Over four days in July, it read: 'Herring gull, pipistrelle, blackbird, sparrow, wood pigeon, wood pigeon, wood pigeon, swift, squirrel, magpie, feral pigeon, feral pigeon, h. hog, bullfinch, swallow, vole, vole, guinea fowl'.

Many farmers see such rescue centres as another misguided form of meddling in the countryside and Secret World certainly provided succour for thugs: there were a crow, a jay, a jackdaw and several grey squirrels being restored to good health. Pauline's roll call of wounded animals offered a fascinating insight into the species that were thriving in the wild and those that were fashionable in captivity. Until 2006, she had never received an otter. Five years on, she had regular admissions, testimony to their recovery in Somerset. She also had a pond of turtles from the Teenage Mutant Ninja Turtle era and currently received far too many pet owls, a legacy of *Harry Potter*.

Like Grandma, Pauline worked on the assumption that, if they could be revived, wild animals were not to be converted into pets. They belonged in the wild. Nevertheless, over the years she had accumulated, like the badger tat, a few exotic species that could not be released (it is against the law, for instance, to return grey squirrels to the wild because they are a non-native species), as well as 'imprinted' wild animals that had morphed into honorary people, with names and personalities – even if, caged, they rather lacked human agency. A cockatoo called Tikka lived in Pauline's small office because no one else would put up with such a noisy old bird. Mr and Mrs Grumpy were a pair of African grey parrots. There was a rescued polecat who was christened Maurice the Mink (a case of mistaken identity) and twenty-seven red-eared sliders (those mutant turtles) in their own bespoke pond. Making up her menagerie were Bernard the turkey and Dennis and Daphne, two tame eagle owls with a diary chock-full of educational talks.

And then there were the badgers. Pauline's badger pens stood in a remote field, away from the bustle of the cockatoos and ferrets, and were arranged with four fox pens surrounding them, preventing any passing wild badger from spreading disease to the recovering patients. Many farmers were furious that animal hospitals could release rescued badgers in any part of the country and were under no obligation first to test them for bovine TB; and this fury seemed rational. Pauline, however, had always vowed to minimise the risk of her rescued badgers infecting others. She and Derek had been dairy farmers, after all, and knew how tough it was. They had had a bovine TB outbreak in their dairy herd after they began badger rehab. 'They said it was the badgers but all three cows were false positives [reacting to the TB test but not actually having the disease],' she said. At Secret World, every badger was tested. If positive, it was put down and the carcass underwent a rigorous post-mortem to confirm whether it was a true positive or not. Of 1,000 badger cub admissions, she had put to sleep about sixty. Of these only fourteen were found to be indisputable positives at post-mortem. 'It shows that bovine TB isn't the problem it's believed to be,' she argued.

All of her uninfected admissions were vaccinated against bovine TB and tattooed on a back leg for identification purposes. As advised by the government scientists who studied badgers at Woodchester Park in the Cotswolds, her cubs were triple-tested, one month apart, to ensure any disease was diagnosed, because bovine TB spreads slowly and can ebb and flow.

Most of Pauline's badgers arrived in April, when cubs first left the security of the sett and became lost, confused or malnourished. 'We get the boys in first because they are a bit thick and the girls come in afterwards,' she said, an observation supported by science: while equal numbers of male and female cubs are born, the badger census at Woodchester found a sex ratio of 39 per cent males to 61 per cent females, showing that male badgers suffer a higher mortality than females.

Earlier that year, Pauline had received her smallest ever admissions: three cubs, all blind, and each the size of a thumb. They weighed 58g, 72g and 76g and were so tiny that their minuscule ears were still sealed. It seemed a miracle they reached Secret World alive, but this was just another example of the badger's resilience, which Pauline experienced all the time. The three cubs came from a family who were renovating a house in the Welsh countryside. One day after torrential rain, they heard squeaking from inside a metal drum that lay on its side on the building site, and found that a sow had moved the cubs from her flooded sett into the drum. The family wrapped the babies in a blanket and left them by the sett, but no animal came to collect them. By now the tiny, chilled cubs looked all but dead. Picked up and placed in a warm car, they revived, and a relay of extremely dedicated volunteers drove them from North Wales to Somerset. They were very small but no longer had their umbilical cords, which suggested they were at least three days old and had therefore received the first milk, or colostrum, from their mother. This, crucially, had boosted their immune systems. When they arrived at Secret World, the three needed feeding day and night, every hour.

Despite having thirty full- and part-time staff, Pauline stayed up all night with them in the kitchen, her warmest room, feeding them baby milk in a bottle. She could not enlist help, she said, because these orphans needed the security of one carer. 'Their eyes and ears are closed so their main sense of security is their sense of smell. And if it's one person they feel secure,' she said. All three cubs survived, grew quickly, and were successfully released that autumn.

Badgers were ideal patients: despite having been bitten once, in all her years of handling virtually every type of British animal, Pauline found them 'very quiet'. Vets, she said, had far more problems with vicious cats. 'Give me a badger any time,' she said, criticising another animal hospital where badgers were handled with graspers and injected through a cage. She believed it was possible to keep a badger calm in any circumstances – if you wanted to administer a vaccine, for instance – by covering it with a blanket. No one chooses to love an animal just because it is easy to handle, however, and Pauline was drawn to them for more emotional reasons. She liked their smell – 'When you think that musk is the basis for nearly all our perfumes it makes sense that you like it' – and she loved that they were elusive and hard to get to know. 'Anything that is secretive is fascinating,' she said.

Most of all, Pauline appreciated the individuality of badgers. With the exception of our pets, which we recognise as unique, ever since Aesop (as Kenneth Grahame realised) we have casually ascribed personalities to whole species: crafty foxes, stupid sheep, vain peacocks. Pauline soon discovered that every badger was different and injured badgers responded very differently to treatment and captivity. 'You can have badgers in the kitchen from birth and you can still release them, or you can have a three-month-old badger from the wild and because you put cream in its eyes for two days it becomes too tame to be released,' she said.

If all went well, as orphaned cubs grew older, Pauline placed them in large enclosures, in groups of five to eight, creating ready-made social groups in which they would ultimately be released. Suitable sites were surveyed to ensure they were empty of badgers and that the release was supported by the landowners, because it was pointless to create a new cete of badgers if they just got shot. Pauline's current release sites were in Norfolk, Lancashire and Cheshire. Rather like Brian May's badgers, her groups were given a soft release: that is, placed on the land with a straw bale sett and surrounded by an electric fence where they would be fed for two or three weeks before being fully freed. The oldest released badger they had found so far was killed on a road: its tattoo showed Pauline had cared for it ten years earlier.

As the controversy grew over the proposed cull, Pauline Kidner, like Brian May, found herself speaking out against it at rallies and on tele-vision. Her hospital was only a few miles from where badgers would be killed. It was possible she would receive new patients – badgers bearing gunshot wounds – when the shooting began. Pauline was alive to the irrationalities in our attitudes towards different animals (one example she gave was how people hated magpies but loved the similarly predatory woodpecker) but she insisted her view on the badger cull was not contradictory. A cull of roe deer, for instance, was acceptable to her because they were non-native species and they could be shot cleanly – shooting deer at night, when the risk of wounding an animal is much greater, was also forbidden. 'It isn't a case of thou shalt not touch badgers,' she said. But she was deeply pessimistic about the prospects of a successful cull; it would be impossible to shoot a group of badgers simultaneously at a feeding station, and so injured or frightened animals would scatter and spread disease. Pauline was not confident she would be able to help wounded badgers, which would go to ground. 'Badgers will do their absolute best to get back to their setts,' she said, 'even if they know they are dying.'

12

The Peanut Eaters

The badger in Trap 007 lowered its head calmly, conserving its energy for whatever tribulations were to come. For the last seven nights it had feasted on a heap of peanuts found inside a wire cage that had suddenly appeared on the Cotswold pasture close to its underground home. On the previous evening, however, when the boar snuffled out this nocturnal treat, the cage door came clanging down. A trap.

Watching this caged badger, I was reminded of a comment made by the irascible old countryman Phil Drabble. 'I see a good many wild creatures under great stress in the course of my work, but I never cease to marvel at the immense dignity of most animals when the pressure is really on,' he wrote. This badger's snout was muddied from pressing into the earth between the mesh in an attempt to escape, but it did not cower when it heard our clumsy approach and refused to flinch when a sharp needle was plunged into its flank. At first, the boar remained motionless when the cage door clanged open. Then, gradually realising that the breeze drifting in was now unimpeded by a

251

door, it made three tentative steps before accelerating away from us in a low dash, as fast as a man could run, ripping through the nettles and under the brambles to the safety of its sett.

The badger cull was looming. Many conservationists argued that the vaccination of badgers against bovine TB could be a realistic alternative, but the government in England was not prepared to fund a complicated and expensive programme whereby the animals would be individually trapped and vaccinated. Gloucestershire Wildlife Trust had taken matters into its own hands and, in 2011, launched on its nature reserves the first ever independent bovine TB vaccination programme for badgers. Field trials found that an injectable vaccine, the Bacille Calmette-Guérin (BCG) vaccine, based on the human version, reduced the risk of badgers testing positive by 54 per cent. The trust was spending £30,000 on its five-year vaccination project.

A red kite soared overhead on a fine August evening at Greystones, a small farm owned by the Trust on land sloping down to the Rivers Eye and Dikler. John Field, the county's mammal recorder (water voles were his speciality), was one of two Gloucestershire Wildlife Trust staffers given the daunting task of trying to catch and vaccinate every badger on Trust property. In the back of his white Mitsubishi truck were the intriguing tools of his trade: a spade, a gun, a portable fridge, a bundle of hay, wire, string, an old army gas-mask bag filled with peanuts and two Useful Sticks.

Badgers had lived on this land long before our Neolithic ancestors cleared the dense woodland and created causewayed enclosures for farming. Greystones had been the site of a hill fort and later the Romans constructed the Fosse Way near by. Gravel was extracted from the valley bottom, forming lakes; cattle were reared between ancient hedgerows. When the Cotswolds became wealthier, farm labourers were replaced by Londoners, barns were converted into homes with gable-roofed garages the size of the old cottages, and the Trust acquired the farm as a nature reserve. As the perpetual restlessness of human affairs played out, the badgers made their homes under the ancient hedgerows, adapting to the constant reshaping of the countryside and enduring the periodic fits of persecution directed against them.

If contemporary agriculture's mix of earthworm-rich pasture and badger-friendly crops such as maize was inadvertently farming badgers, then Gloucestershire Wildlife Trust were doing it rather more consciously. Labyrinthine setts were allowed to sprawl from hedgerow into field; meadows, untreated by pesticides, provided perfect nocturnal foraging grounds. But for all its encouragement of the wild flora and fauna on its land, the Trust was also a farmer, and it was tending a prize herd of Friesians. There were eighteen in calf and the Trust wanted to keep them bovine TB-free because it planned to make single Gloucester cheese with a local producer, a potentially profitable venture. If these animals contracted the disease the farm would be 'shut down' and every cow that reacted to the TB test would be slaughtered. The charity needed to protect its cattle as well as the badgers.

The biggest argument against trapping and injecting badgers was its expense. It was expensive because it was arduous. John Field, an athletic-looking rugby fan who looked more like a conventional farmer than a conservationist, began by placing twenty traps, sturdy wire cages big enough to hold a dog, close to the three setts on the farm. Each trap had to be dug into the earth because the sensation of the metal cage beneath their feet was enough to dissuade badgers from entering. For seven nights, John 'pre-baited' the unset traps with peanuts so that the badgers became accustomed to popping in and out for a nightly scoff. 'We're not talking rocket science here,' he said.

After all this preparation, tonight, finally, we would set the traps to catch them. We bounced down wide field margins of beige grasses, fast-fading thistle flowers and tatty Meadow Brown butterflies, all bleached by the sun. On the neighbouring farm, an orthodox, industrialised venture, everything was scaled up and a hulking blue tractor and a red bailer (why does farm machinery always come in primary colours?) were whizzing around an enormous field.

By each group of traps we stopped the truck to reload them with peanuts. Then we tied twine from the swing door to a piece of wire that was looped around a heavy stone we placed inside the cage over the nuts, which prevented them being eaten by smaller mammals, such as squirrels or mice. We hoped badgers would enter the cage and nuzzle the stone to get to the peanuts, thereby tugging the wire and string, which would bring the cage door clanging down. 'It does strike me as funny that despite today's technology we're catching a badger with a bit of twine,' said John. 'Yes it's labour-intensive but the "does it work?" question has been answered. The vaccine works. Otherwise it wouldn't have been licensed for use.'

As we set the traps, we chatted about this and that. When he was a child, John had been inspired by one book, *The Amateur Naturalist* by Gerald and Lee Durrell, and followed its instructions to reconstruct a hedgehog skeleton. 'As a kid I wanted to be a naturalist or an Egyptologist or a vulcanologist, or a sniper padre because I liked *The Longest Day*. Not being particularly good at physics [or killing] I went down the naturalist route,' he said. He was well qualified to catch badgers: after studying zoology at university, his first job was as a research assistant on the Randomised Badger Culling Trial, studying biosecurity on farms. We talked about my book and I told him something I had not dared mention to Judy Salisbury and some of the other badger lovers: that I wanted to eat badger as part of my exploration of our relationship with the species. It turned out that when John was working on the culling trial, he had met an old woman in Cornwall. 'Ooh, badger ham,' she said. 'Handsome that is, handsome.'

Wrestling with the task in hand, I was struggling to get the string appropriately taut. I was also worried about spreading my scent all over the traps. Badgers are difficult to trap or snare because they will shy away from strange implements that smell of human. The naturalist F. Howard Lancum once placed his handprint very deliberately on a badger path at 3.30 in the afternoon, waited, and found it caused a badger to bolt down its sett six and a half hours later when it picked up his scent from the patch he had 'marked'. Surely my stench would scare off the badgers? John was not concerned. They did not mind his scent and had been taking the bait all week. Besides, the badgers around here were used to strange goings-on: government scientists were testing oral badger baits up the road. But what about accidentally trapping foxes and otters? 'No otter in their right mind is going to eat peanuts, but we can't have that on my watch. That's why there aren't any traps by the river,' said John. Foxes were occasionally trapped, and released. And dogs? This was a risk and there were warning signs for dog walkers across the reserve. It was another reason why the traps had to be checked very early the next morning. We retired for a quiet pint, one eye on the time. We would rise again at 4 a.m. to see what we had caught.

Vaccination was every anti-culler's favourite word – and no wonder, because it was an attractive panacea: a scientific fix that avoided the slaughter of badgers. Every luminary I met, from Brian May to Sir David Attenborough, proposed vaccination, not just of badgers but of cattle as well.

Unfortunately, it was not that simple. Before we set the traps, I met Gordon McGlone, the chief executive of Gloucestershire Wildlife Trust, in the barn on Greystones Farm that had been converted into offices. A pragmatic man with oversized glasses and silver hair, McGlone had arrived at the charity as a young Wildlife officer when badgers were gassed, in the early 1980s, and watched the debate rumble on for decades. 'Leadership, that's what's been missing. It hasn't been grasped in the way that other industries might have done,' he believed. 'There have been lots of culling exercises. All of them were the product of "let's have a go at this" and all were dropped because they made the problem worse. Defra says doing nothing is not an option, but before them the Ministry of Agriculture and the Ministry of Agriculture, Food and Fisheries did a lot, and it made things worse. Where has the learning come from?'

McGlone knew the injectable vaccine would be time-consuming, expensive and partial but was still determined enough to try it. The National Trust also started a five-year vaccination programme on its Killerton Estate in Devon later that year, and by 2012 eleven Wildlife Trusts, including Shropshire, Somerset and South and West Wales, had launched badger vaccination schemes. Defra was also funding a vaccine deployment project over a hundred square kilometres of Gloucestershire to train 'lay' badger vaccinators: in 2010, over 500 badgers were vaccinated; in 2011, more than 600. Most encouragingly of all for the vaccinators – and to the delight of Brian May – the government in Wales decided to vaccinate rather than cull. In 2012, the first year of its five-year programme, 1,424 badgers were trapped and vaccinated 'without incident or injury', according to the Welsh administration, which planned to extend the project beyond its focus in North Pembrokeshire. Farmers, however, were critical of the £943,000 cost of the Welsh operation, which worked out at £3,912 per square kilometre – or £662 per badger, well above Defra's estimate that trap-and-vaccinate would cost £2,250 per square kilometre.

Vaccinating badgers was greeted with scepticism by most farmers and by government scientists in Whitehall. Apart from the expense, it had several flaws: you need to vaccinate 70 per cent of the animals to create 'herd immunity' and ensure there are so few carriers within social groups that the disease will not take hold; there is also no way of knowing, year on year, whether a badger has been vaccinated or not, unless you tag ears or undertake a blood test, neither of which is desirable on wild animals. This meant that badgers would be needlessly repeatedly immunised. 'It's just not cost-effective,' said Jan Rowe, a Cotswold dairy farmer I met, who estimated bovine TB had cost him more than £300,000 over twenty-five years. Rowe had become a leading player in the badger-cull campaign. Although the government promised £250,000 to help farmers vaccinate, this money would barely cover the inoculation of badgers on 100 square kilometres of England.

Rowe was not alone in pointing out that the BCG vaccine was relatively unproven and would take too long. Nor was there any scientific proof yet that the injectable vaccine for badgers actually reduced the level of bovine TB among cattle, although if farmers claimed badgers were a significant transmitter of the disease, then removing it from badgers would logically reduce it in cows over time.

But the scientific evidence for the effectiveness of the badger vaccine was encouraging. The first field trial saw a 73.8 per cent success rate attributed to the injectable BCG vaccine. In 2012, a four-year study led by Stephen Carter in which badgers in a highly infected part of Gloucestershire were vaccinated reported a more modest 54 per cent reduction in the risk of a badger testing positive. Strikingly, however, Carter's research found that when unvaccinated cubs lived in a social group where just a third of badgers had been given the BCG vaccine, the risk that the cubs would test positive for bovine TB was reduced by 79 per cent. This finding was a powerful argument in favour of BCG: it suggested that it would relatively quickly establish something approaching 'herd immunity' among badgers, although the study's authors cautioned that their findings had been obtained over a relatively short time-span.

Nigel Gibbens, the government's chief vet, understood the challenges of vaccine development better than most, and he was cautious about the efficacy of the BCG vaccine. 'We've got limited experience of its use in the field,' he said. An injectable vaccine was likely to be a slower way to tackle bovine TB in cattle than a badger cull because it would not work on infected animals, although research published in 2011 showed it reduced the severity of the disease in badgers and cut down on their excretion of bacteria – potentially making them less infectious. If an average badger lived for five years in the wild, then it would take at least that long to banish the disease from badgers. Given that scientific studies have found that, on average, about a quarter of all social groups contain a badger aged seven or more, and the oldest badger recorded in Wytham was an astonishing sixteen years old at the time of its death, would a five-year vaccination programme really cover all badgers and establish herd immunity?

Trapping and injecting every badger in bovine TB hotspots would be a challenging and cumbersome feat. An oral vaccine that could be smuggled inside peanuts looked like the real solution: simpler to administer, cheaper, and much more effective. Rabies was basically eradicated from Western Europe when an oral bait was developed for wild foxes in the 1980s. In 2011, Defra said it was investing £20 million in vaccine development, with a target of producing a licensed oral badger vaccine by 2015; but everyone in Whitehall was pessimistic about meeting this date. The target for the oral vaccine looked 'increasingly challenging', said Nigel Gibbens in 2011. 'The most recent results were very disappointing. It didn't work. Because of that setback that makes us very uncertain about the time [2015].'

There Is No Alternative was the mantra at Defra. But from what I observed, there did not seem much political will inside the coalition government to perfect an oral vaccine. Six badger vaccination trials were planned by the previous administration but in 2010 the coalition, seeking to reduce public spending, cut five of them. Government spending on developing a vaccine steadily fell, from £3.2 million in 2009/10 to £2.1 million in 2012/13 and a projected £312,482 in 2015/16.

'It's a dirty business. It's become deeply politicised,' was the verdict of Chris Cheeseman, the badger biologist. Cheeseman questioned the negative spin being put on the search for an oral badger vaccine. When the government announced it had problems with the 2015 deadline, scientists 'were totally shocked,' said Cheeseman, who kept in touch with former colleagues developing the oral vaccine at Woodchester Park. Gibbens's assessment was 'completely out of kilter with their understanding of their research and progress', believed Cheeseman. 'They have not reported any significant unexpected issues or unavoidable problems. The things mentioned in terms of delaying the original vaccination – selective bait uptake, dosage rates, survival of the vaccine at ambient temperatures, survival of the vaccine in the gut – none of these were unanticipated.'

Cheeseman was certain about one thing: badgers were easy to bait. 'They love peanuts and they love honey. Mix those together and they find it irresistible.' The bigger challenge was to find a fatty formulation to surround the oral vaccine that could survive in a badger's stomach, enabling the vaccine to be absorbed in the gut. Researchers in New Zealand had developed an 'encapsulation technique' for a bovine TB vaccine. This had been very effective on possums there, which carry bovine TB. But there were two further hitches. The possum was a herbivore whereas the badger had the more acidic stomach of a carnivore; what worked on possum might not work on badgers. Another difficulty was a human one: the New Zealand researchers wanted to make money from their discovery and would charge the British government for their vaccine. This would make an oral badger vaccine a more expensive solution.

A soft drizzle was falling at 4.58 a.m. as we bumped back down the field margin in John Field's pickup to check the first trap. We pulled up in the corner by a discarded bottle of Smirnoff Ice and a tennis ball and John cut the engine. In the pre-dawn silence, the only noise was the fridge in the back, which contained 1-millilitre measures of diluent and the BCG drug. Each £15 shot of vaccine had only a four-hour lifespan and so had to be prepared in the field. If we found a badger, John would inject the diluent into the vaccine and then draw the mixture into the syringe. Every animal got its own needle; used needles were dropped in the yellow sharps box. The vaccination programme was closely monitored: the traps could only be operated under licence from Natural England, the government's wildlife quango; trapping could occur only on two consecutive nights because three would deprive too many badgers of watering opportunities and recaptures could stress the animals; most importantly of all, every animal had to be freed from its cage by 8 a.m., to minimise its suffering. Optimistically, John predicted his twenty traps would have caught twelve badgers. If we had caught this many – I was doubtful, having set some of the traps myself – it would be a race to vaccinate them all before 8 a.m.

We climbed out of the truck and John slipped on his protective glasses (purchased from an ex-soldier flogging his personal protective gear on eBay) and mask. He wore these because a nervous badger might sneeze or spit in his face, and it would certainly be a PR disaster for the anti-cull movement if a vaccinator succumbed to bovine TB. 'Cubs try and have a fight with you on their way out. Sometimes they hiss and spit a bit,' said John. 'But the greatest health and safety issue is the fact that it's a farm, it's dark and there are barbed-wire fences which could snag you.' One final piece of equipment remained safely stowed: the gun, in case we accidentally caught a grey squirrel or mink. It is against the law to release non-native animals if they are caught in a trap, so John would have to shoot them. Lacking the required veterinary qualification to inject badgers, my job was less heroic: I was a pre-dawn pen-pusher, clipboard in hand, tasked with 'pharmacovigilance' – recording the health of each animal before and after its injection, and the time of its release.

Rain tapped gently on the pasture and a cockerel crowed. Swish, swish went our waterproof trousers as we walked across the wet field. From over by the hedge, obscured from view, came a banging sound, a badger clattering against a cage. With a snap, John stretched a pair of blue plastic gloves over his fingers.

The first trap was empty, the gate still up. The second trap was sprung – the gate slammed shut – but mysteriously vacant. In the two minutes it had taken to check the first pair of traps, the banging had stopped. There, curled in the third cage, was the source of the noise. But this grey blob did not move. We peered over the motionless badger. John clicked his fingers by the creature's ear. Nothing. It could not be asleep, so soundly, so soon after making that banging. Was it dead? Had it suffered a heart attack?

Unperturbed, John noisily prepared the vaccine and rattled the cage. Finally, with a flick of a small ear and a twitch of nose, a button eye opened, a female badger rudely awakened from sleep. John leaned in to plunge the needle into her rump through the wire cage. The badger sat there, very Zen. 'Vaccination done,' announced John. The badger had not even flinched. As she quietly put her nose to the cage door, a patch of fur was clipped and sprayed with red stock-marker, a temporary mark that would show which badgers had already been vaccinated when the traps were checked again the next morning. John opened the gate and the badger stepped out with the caution of an animal weighing up whether this unexpected development spelled escape or doom. As soon as she stepped clear of the cage, she shot off, heavy-footed but surprisingly fast, thundering into a hole further along the hedge line. Sometimes, said John, the badgers were so unfussed by their confinement that when he released them they paused to hoover up any peanuts they had missed earlier.

The young badger with a long scraggy tail we caught in the next cage was not so insouciant. The first had been immaculately groomed, reminding me of a Victorian hunter's observation that a bedraggled badger baited in a pub yard would disappear into its box for a few seconds and emerge with no hair out of place, finding time to meticulously groom itself even while being terrorised by a dog. But this youngster was filthy, covered in soil from attempting to dig out of the cage. For at least a foot around it, the earth had been picked clean by the badger's surprisingly long forearms, which had stretched through the mesh, torn at the grass all around the cage and dragged it into the wire box to create a nest. When we approached, the badger shifted, as easily as a marble, the heavy stone that had been placed over the peanuts inside the cage; and then, head down, displayed its badge in warning – a little bull, ready to charge. John struggled to hold it still as it leapt about, turning quickly whenever he tried to inject its rear. 'I've got to get his back end, that's the only place I can inject,' he grunted. Finally, the grubby adolescent was cornered and jabbed.

Courage is widely seen as an attribute of the species, and most badgers greeted us, a nightmarish four-legged apparition bearing big needle and clipboard, with an equanimity that could be stoicism or terror, or both. But I saw for the first time how differently individual badgers responded to a stressful predicament. In contrast to the stroppy young badger, the second juvenile we injected stopped and sniffed John's leg like an inquisitive dog when it was freed. Another, which had dragged an old silage bag from who knows where into the cage, ran short-sightedly straight at me when it was released, pushing my legs apart like an enthusiastic puppy – my first physical contact with a badger. A third demonstrated the most severe case of Stockholm syndrome: it had rustled up a lovely nest of dry grass inside the cage and for several minutes refused to leave its new home.

Others seemed more overwrought. An attractive badger with very fine stripes emitted a low groan like the mooing of a cow. She gave off a strong smell, halfway between wet dog and damp horse, with an added gamey note, like the fug of the nocturnal animals' house at London Zoo. Equally apprehensive was a big young badger who snorted and sneezed and bounced around in the cage for several minutes before it could be gently held still and injected. When it was released it darted through a tiny hole in the fence. Despite some signs of stress, John was not worried that we were causing irreversible trauma. 'Most of them were asleep until we've got there. They have a stress point when they are captured but after a while they calm down. It's one night in a lifetime, it's not that stressful in comparison with everything else they meet.'

We hit John's prediction of twelve badgers, with three traps to go. 'All for your benefit, cows,' he said to the herd watching lugubriously from the gate. Shortly before 8 a.m. we reached the final trap, under the willow trees in the valley bottom. Here, wet and dirty, was the fourteenth badger of the morning. Watching trapped badgers close up left me with three firm impressions: how calm they were under fire, how heavy-footed they were when they ran away, and how difficult it would be to cleanly kill such a fast-moving animal in the dark.

By the end of the first year of its programme, Gloucestershire Wildlife Trust had vaccinated forty-two badgers on seven nature reserves. It planned to continue every July for the following four years until, it hoped, old badgers bearing bovine TB would have died off and herd immunity had been established. This was vaccination on a modest scale, but after catching many of these animals and plunging the syringe into their rumps John Field was convinced it could be done more widely, by farmers themselves. The problem was not catching the badgers, it was luring the farmers. Gordon McGlone's hope that the Trust's vaccination policy might break down the barriers between conservation and farming appeared to be in vain. 'To us this is an industrial problem,' said McGlone. 'This isn't something fringe, it isn't just about badger conservation, it's a major industry which Gloucestershire Wildlife Trust are part of. We've been shut down by bovine TB here. We have felt the direct effects of the disease.'

So far, however, despite an open invitation, no farmers had visited Greystones Farm to see how its vaccination programme was working. 'It's such a polarised debate. Badgers are wonderful/badgers are evil. Neither of these is true,' sighed McGlone. 'They are a land mammal caught in the middle of a complex ecology and the farming industry.'

*

Apart from the injectable and oral badger vaccines, there was a third vaccine proffered as a solution to the problem of bovine TB: a vaccine for cattle. By 2012, this one, based closely on the BCG vaccine, was ready to be licensed. It was already being deployed in countries such as the USA, Argentina, Ethiopia, Mexico and New Zealand. Small-scale field trials in Ethiopia, over two years, introduced both vaccinated and unvaccinated young cows into a herd where every animal was infected: 56 per cent of the vaccinated animals showed no signs of infection compared with 100 per cent of the non-vaccinated ones. A similar BCG trial in Mexico found its success rate to be 60 per cent. The vaccine's immunity lasted for between one and two years, so cattle would have to be vaccinated every year; the cost of one shot was estimated at anything from £8 to £20, so it would be expensive.

The more immediate problem was that vaccinated cattle would react to the tuberculin skin test as if they had bovine TB. So another test, to distinguish between a vaccinated cow and a cow with TB, was required. In the government's Animal Health and Veterinary Laboratories Agency (AHVLA) labs at Weybridge in Surrey, researchers came up with the memorably named Diva – differentiation of infected from vaccinated animals – blood test. Like the vaccine, this test was ready to be deployed at the end of 2012, but there remained one obstacle: the European Union. The vaccination of cattle against TB had been prohibited in the EU since 1977 because of the uncertainty it created over the bovine TB skin test. Until the EU ban was lifted, a licence for the vaccine could not be issued. The Diva test would remove the need for a ban on the vaccine but Diva could not be licensed until the vaccine itself was made legal. This was a tangle of political dimensions.

A vaccine for cows with a 50 per cent effectiveness might not be a panacea but it would be a good start. If we were actually a tweak of EU regulations away from halving the number of cows contracting bovine TB, then surely a spell of sustained lobbying by British ministers could change the European laws? It did not appear, however, that Defra or the prime minister had been exerting much pressure in Europe – which was strange, given the tough Tory rhetoric over Brussels bureaucracy, sovereignty and red tape. In an inspired piece of campaigning, Brian May spotted this discrepancy and organised a trip with Gavin Grant, chief executive of the RSPCA, to meet EU officials and discuss the obstacles to a cattle vaccine. Even more smartly, May invited along the NFU, who dispatched a representative.

In an account of his trip to Europe published in the *Mail on Sunday*, May described how he had asked Georg Haeusler, chief adviser to the European Commissioner for Agriculture, why the EU would not allow Britain to vaccinate cows. 'He looked at us in surprise and said: "But this is not true. You British are welcome to. You would find it was not possible to sell cows into the mainland of Europe because we would be risking bringing bovine TB into our countries. But you do not export live cows to us anyway. It would be meat and milk and other "products" made from cattle that would be proscribed. But there would be no police descending on you if you began vacci- nating tomorrow.' Haeusler also told May that if Britain could prove its Diva test worked, 'There would be no reason for us to ban the import of the products derived from your cattle.' Although May left the discussions believing the EU ban on a cattle vaccine could be quickly changed, the European Commission subsequently issued a statement that warned it 'would take time' to get the Diva test approved at 'EU and international levels'.

Pro-cullers continued to play down the efficacy of a vaccine for cows. Farmers argued the reservoir of TB in wildlife would still need to be tackled, otherwise cattle would continually get reinfected. Furthermore, even if researchers developed a more effective version of the vaccine, which was likely, scientists said it would not miraculously cause TB to vanish. 'I don't think it can be a strategy on its own – you have to use all the tools in your toolbox,' said Glyn Hewinson, chief scientist at AHVLA. 'Everybody agrees that vaccination is the only long-term solution but what people don't realise with vaccination is how long it takes to be effective,' Tim Roper told me, citing Defra models for the vaccination of badgers which predict it could take forty years for bovine TB to be eradicated. Whether we vaccinated badgers or culled them, the disease would continue to infect our cattle for many years to come.

13

Lunch

Behind the curtains of a bland suburban semi was a very unusual living room. A stuffed green woodpecker perched by a dim lamp in the far corner; opposite stood a buzzard and a short-eared owl; behind them, a tawny owl, aloof in its perpetual motionlessness. Joining this strange flock were a fruit bat hanging from a bookcase and a wild cat snarling over an armchair. The picture rail carried dozens of sets of antlers. Below them were the mounted heads of a Himalayan sheep and a four-horned Hebridean sheep. Slung over the back of the chair by the table was a ladies' jacket cropped from the skin of a snow leopard. On the floor were three antique microscopes, beside which were assorted samples of animal scat lined up to be placed beneath their gaze.

The warm air of the first-floor flat at the dreary end of Bournemouth smelt strongly of mothballs. Beyond the pungent scent of paradichlorobenzene was the whiff of something else: decay. I knew what I had come here for so I should not have jumped when I popped to the toilet and found the carcass of a badger slumped in Jonathan McGowan's bath. As heavy as hand luggage and smelling mostly of wet dog, here lay lunch.

I had been pursuing Jonathan for some weeks. Ever since I'd discussed badger meat with John Field, I knew I must try it. I had met a pensioner in a pub in Exmoor who exclaimed, 'Badger 'ams, best 'ams I ever tasted', but people of my generation were not known for their strong stomachs. My mum used to pull over and pick up dead pheasants without batting an eyelid but I had never butchered any kind of animal and lacked the nerve to collect and carve up my own roadkill badger. Instead, I had stumbled across Jonathan while researching a story about moves to reintroduce the lynx into Britain. He wrote a fascinating blog about feral big cats, posting his pictures of evidence of footprints, scats and the carcasses of their prey animals. There was something of the Victorian amateur naturalist about him, because as well as being an expert tracker he was a taxidermist and an advocate of eating roadkill. 'Owl curry, adder with butter and stir-fried craneflies!' squealed the *Daily Mail* with characteristic brio when they interviewed Jonathan for a story about how this '44-year-old bachelor' had supposedly survived on a diet of roadkill for thirty years. Here, behind the belittling tabloid jollity, was a man brave enough to eschew the bland tastes of mainstream society.

When I phoned and explained I wanted to eat badger, Jonathan agreed as if it was as natural as requesting a pint. Our first date was cancelled because he could not find a roadside carcass in time, but a week before our second proposed lunch, he picked up a nice fresh badger on a main road by a nature reserve close to the Hampshire–Dorset border. I admired the species and saw no contradiction between liking an animal and wanting to eat it. Jonathan agreed. 'I'm a wildlife lover and expert and conservationist but I also like to eat them,' he said. 'Other conservationists say, "How can you do that?" Well, why not?' Roadkill badgers were the accidentally slaughtered, free-range waste products of our excessively paced, car-based society – surely the most ethical meat feast of all.

'Shall I bring some hot pepper sauce?' I asked when we arranged lunch over the phone.

'You can if you like,' Jonathan replied.

'Why are we exterminating badgers and not cows over bovine TB?' asked one dissenter, rhetorically, regarding the cull. 'Because cows are good to eat and badgers taste horrible.' Badger has long been repugnant to mainstream palates in Britain but I was not yet sure if this was because of a taboo against eating the animal or because its rich, dark flesh was disgusting. The sentimental love of animals that flourished in the nineteenth century caused many species to be banished from British menus. A not strictly logical menagerie of household pets and wild animals, including dogs, cats, horses, badgers and songbirds, were all deemed beyond the pale. 'What! Robins! Our household birds! I would as soon eat a child,' exclaimed the historian Mountstuart Elphinstone, aghast at witnessing the consumption of songbirds when travelling in Italy in the nineteenth century. Kenneth Grahame cannily ensured that none of his four heroes in *The Wind in the Willows* were animals we would usually eat.

Badger is absent from the great cookbooks of the past, and the attitude of the American food writer Waverley Root was typical: he dismissed the badger as food for eighteenth-century English peasants, who would buy something more succulent if they could only afford it. There has, however, always been a hidden country tradition of *Meles meles* meals. In 1774, the Scottish author Dr John Campbell wrote: 'The Badger is hunted and destroyed whenever found and being by nature an inactive and indolent Creature, is commonly fat, and therefore they make his hind Quarters into Hames in North Britain and Wales.' Over the centuries, a few cooks attempted to persuade sceptical readers of the merits of the badger and, in particular, its hams. Jacob Robinson and Sidney Gilpin, authors of a late-nineteenth-century book about North Country sports, said badger hams 'were esteemed superior in delicacy of flavour to the domestic pig or wild hog. In this country, the hind quarters only were used for food; while in some parts of Europe and in China, the whole carcass was held in high esteem, and considered to be very nutritious.' Gypsies once begged the naturalist J. Fairfax Blakeborough for the body of a badger he had trapped in Yorkshire, telling him it tasted far better than the pale flesh of hedgehog.

The badger-ham-loving pensioner I had met on Exmoor was not a complete anachronism, either. When the badger scientist Chris

Cheeseman moved to the Cotswolds in 1975 he got chatting one night in the pub to a local who treated his rheumatism with badger fat. 'He said, "Badger fat is wonderful – it will sweat through glass",' remembered Cheeseman. He was invited to the local's cottage to try badger. The cured hams looked like small legs of pork. 'It was a bit like very dense, dark-red ham,' Cheeseman told me. 'You'd have to say it was very tasty.'

Henry Smith's *Master Book of Poultry and Game*, written for the catering trade in the 1950s, included instructions for curing and baking badger hams, making a pie from its forequarters, roasting the legs with a seasoning of ginger and an accompanying sauce of horse-radish and gooseberry, and even using its feet and tail for gravy. An annual badger dinner was held in Castle Cary, Somerset, every year until 1965; during that decade badger diggers were offered £5 for a pair of badger hams in Stoke-on-Trent, a decent sum of money which suggested there was still demand for prime cuts. A few hotels and restaurants had badger hams on their menus as recently as the early 1970s, before the creatures were given legal protection. Clarissa Dickson Wright, the celebrity chef, may have been exaggerating when she claimed badger hams were for sale at the bars of most West Country pubs when she was a teenager – 'just like a jamón ibérico' – but it was interesting to see the horrified reaction to her suggestion, on the eve of the cull, that the carcasses of the shot badgers should be eaten instead of being bagged up and sent for incineration, as planned. Dickson Wright, who described 'delicious' badger hams as tasting rather like young wild boar, said badger had been a popular staple in the Middle Ages, when any kind of meat was considered a luxury.

Continental attitudes towards badger on the menu have always been more relaxed. 'In Italy they eat the flesh of Badgers, and so they do in Germany, boiling it with Pears: some have eaten it here in England, but like it not, being of a sweet, rankish taste,' wrote Nicholas Cox in 1677. In France, badger is not a mainstream dish, although *blaireau au sang*, 'badger with blood', is a traditional recipe, and there may be more historic enthusiasm for the meat. During the First World War, the badger naturalist Mortimer Batten and his French colleagues disturbed a badger while riding a motorbike and sidecar along a lane three miles from the front. The French soldiers urged the English badger-lover to run it down because, they said, *blaireau* was excellent to eat. There is also a long Alpine tradition of eating badger. Mortimer Batten recorded that badger ham was 'said to be excellent' in Italy, where it was likened to bear meat; and badgers are still legally shot in Switzerland, where hunters consider the meat a delicacy. The badger scientist Tim Roper consumed it in the Alps. 'I found it tasted of the forest from which it came, with flavours reminiscent of earth, leaf mould, pine needles and fungi. In any case, it was not very palatable,' he wrote. Further east, badger is still regarded as a peasant food but it is used in goulash in remote parts of the Balkans, and in shish kebabs and sausages in rural Russia.

In the 1980s, the cookery writer Tom Jaine gave a detailed account of his badger meal. He took a hindquarter from a badger accidentally caught in a fox trap and marinated it in red wine, bay leaf, thyme and parsley, carrots and celery . . . for a week! After browning it, he pot-roasted the joint with the marinade in a cool oven for three hours. 'We found that the most useful comparison was to mutton. The meat was dark, succulent and strong tasting, but in no way like pork, having a particular smell to it,' he wrote. 'The fat, of which there was ample, was not much enjoyed.' Did Jaine savour the flavour of his badger? 'I do not think we did. The taste was by no means unpleas-ant, although rich. However, the psychological difficulties in eating a truly wild animal weighed heavily upon us. Had we thought it some variant breed of lamb, delivered by the Rare Breeds Farm, we would have been interested and mildly enthusiastic. None of us has the stomach to consume wild things.'

I was surprised by Jaine's feelings. Of course, our society does not find it acceptable to eat rare wild animals. But what, apart from taste, makes a serving of common badger so psychologically troubling? Historically, our rejection of songbird pies, for instance, was due to increasing prosperity; perhaps, if society collapsed, we would return to eating them.

This was the view of Angus Wilson in his 1961 novel, *The Old Men at the Zoo*. Set in a dystopian near-future, when a totalitarian European movement is at war with England, it tells the story of Simon Carter, the secretary of London Zoo, who struggles to save its exotic animals from a starving mob. Carter flees to rural Essex with a lorryload of lemurs, tarsiers, pottos, lorises and a dead gorilla, but is ambushed by villagers and finds a haven from anarchy in a wood, where he watches a boar badger, its mate and two cubs playing. This 'healing innocence' is rudely interrupted when a shotgun-toting teenager kills the boar and one cub. Carter is taken in by the teenager and his mother and served 'a plate of what appeared to be pinkish greasy, fried pork'. To his surprise, the badger tasted 'rich and delicious', he recounts. 'In a short while I had eaten all but a small piece of browned fat. I speared it with the fork, felt its grease against my lips, and then suddenly I vomited. So violent were the spasms that it seemed as though my body were rejecting all its vital organs. I spewed a flash of light vermilion blood. The room span round. My head fell back on the cushions and everything became dark, became nothing.'

In the next chapter, the war has ended and Carter is contemplating a diet of 'juicy steaks and red wine', sitting around a table discussing 'ancient European culture, international science, sound economy and civilised living' – everything, in short, that badger eating is not. Just as Carter collapses with dysentery, the war ends; and when he recovers he is put in charge of London Zoo again. This is not the happy ending it might appear, however: a new European government is in charge, habeas corpus is suspended, and a mangy brown bear is chained up inside London Zoo, a sign that the new regime advocates the pitting of wild beasts against political prisoners as a public spectacle.

In Wilson's dystopia, when the British love for animals is subsumed by other more desperate priorities, society as we know it collapses, along with freedom and democracy.

LUNCH

Consuming unusual animals is one short step away from eating each other. Just as the mob resorted to eating the rarities of the Zoo so the respectable, animal-loving Zoo secretary devoured innocent badgers in a wood, and both crossed some kind of Rubicon.

Slightly awkwardly, Jonathan McGowan and I squeezed into his small bathroom and prodded our lunch, which was defrosting in the bath. A medium-sized male, about three years old, its shoulder blades were firm beneath its thick skin. Its muscles still felt ready for action but the narrow head was completely spongy, its skull crushed, because this badger had turned to meet its quarry, the motor vehicle, head on.

This was characteristic roadside behaviour. A *Manchester Guardian* 'Country diary' column of 1931 told a fascinating story of a young lad who was driving to a car mechanic's along the dark lanes of Cornwall when he ran over a large object. At that moment, 'a big animal leapt on to the car and made for the driver most fiercely. The driver, terrified, stamped on the accelerator, shook off the creature, and safely reached the garage.' The mechanic scarcely believed the story but there was something about the lad's terror that persuaded him to inspect the site of the accident the next day, where he found a large dead 'dog' badger. They concluded that the ferocious wraith had in fact been Mrs Badger, who was 'inspired by rage to magnificent courage' and 'flung herself upon the unknown monster to avenge her mate', as the columnist, W.A.F., put it. Perhaps the motor car brought badgers into focus, and their slaughter made them objects of sympathy.

281

First, butcher your badger. This task intimidated me but was 'common sense', according to Jonathan, who knelt down beside the bath. 'An animal doesn't need to be taught how to eat an animal but we somehow need to be,' he said, taking a scalpel in his hand and deftly unzipping the thick hide and exposing a yellowy outer layer of fat. 'I'm going to cut along here, take out one side of the ham and just use a little bit of it.' I love fatty meat, like belly pork, but according to Jonathan, badger is different: its fat is mostly outside its flesh, distributed very thinly in lines throughout its muscle sections. He delved into the badger's left-hand hind leg with his scalpel. Fatty skeins were pulled away as he carved small chunks from its body. The flesh was dark vermilion and attractive, those veins of fat no more than a pencil squiggle running through it. 'Deer have lovely back meat. Badgers have hardly any,' he said as he worked. 'Most of its power comes from its front and hind legs and its neck so badger meat is found on the back legs.' There are few more eloquent testimonies to the lifestyle of a wild animal than its flesh.

Jonathan washed the meat under the kitchen tap to remove a few remaining bristly, stripy badger hairs. His kitchen might not pass a hygiene inspection – a film of grime covered most surfaces and tea bags were dumped on the worktop after use – but Jonathan mocked popular fears of catching diseases from eating roadkill. 'People think wild animals have diseases. We have more diseases and bacteria in abattoirs and farms. I never buy meat from supermarkets,' he declared.

Badgers, however, were known to carry parasites and some risk of disease. In 1835, a year after completing an enthusiastic natural history of the wildlife – including badgers – found near his hometown of Great Yarmouth, James Paget was conducting a human autopsy at St Bartholomew's Hospital in London as a first-year medical student when he noticed minute worms lining the diaphragm of the cadaver. Alongside his supervisor Sir Richard Owen, Paget, who would become a celebrated Victorian surgeon, discovered the roundworm that causes trichinosis, a parasitic disease. It is most commonly caught by eating undercooked pork, but outbreaks of trichinosis in 2005 in the remote Russian region of Altai, close to the border with Kazakhstan, were blamed on badger meat in shish kebabs. Some of the twenty-five infected patients told doctors they ate badger because it was tasty and, more importantly, cheap.

Any risk of trichinosis could be removed by cooking badger as thoroughly as pork, and a good heat would kill any bovine TB as well. 'Even if we were to play around with a dead badger TB is hard to pick up,' reckoned Jonathan. 'You won't get it from the flesh, you'll get it from saliva or blood or faeces or urine. I never worry about TB. Despite all the badgers I've watched and picked up dead I've never caught it.'

Having said that, Jonathan was not particularly fond of eating them. 'They have a particular badgery smell and it taints their flesh. You can't get rid of it,' he said. Others who have eaten badger have compared it to muddy wild boar and the smell of swan poo. Where would badger fall in his roadkill top ten? 'About twentieth. I prefer hares, deer, duck, squirrels, rabbits,' he said. 'Fox is lovely. Fox I rate quite highly – number five, to be honest. Sometimes a fox cub is absolutely delicious.' Grey squirrels were also 'absolutely delicious'. Occasionally, he would eat roadkill not merely rare but raw. Raw deer, drained of blood, was fluffy like cooked apple. 'It melted in my mouth,' he said, remembering the day he tried it. 'It was all nutty and delicious. I thought, now I know what the leopard likes when it gets raw deer in its mouth.' If badgers were that good, he reasoned, they'd be on menus everywhere.

Ignoring all the elaborate historic recipes to make badger more palatable, Jonathan planned to serve a simple dish: badger stir-fry. He skilfully cut the already thin badger steaks into wafer-thin slices, then laid them in a small frying pan with sunflower oil. Frying, he hoped, would remove the peculiar, pungent taste of badger, which roasting can lock in. And as they began to sizzle, sure enough, the slices released a strong odour.

'There's no knowing exactly what it's going to taste like,' said Jonathan. 'Usually it smells of badger but this one smells just like liver.' Perhaps the defining feature of wild meat is its unpredictability. Much depends on the diet of the specific animal, and badgers have a very varied diet. Perhaps we would have to farm corn-fed badgers to get them to taste nice. Or we could accept, as people like Jonathan did, that there were rarely absolutes in the wild, or in gastronomy, and certainly not in wild gastronomy, except perhaps for the principle that the younger an animal, the better it tastes.

He pressed the meat flatter with a spatula, cut off a couple of slices, slung it on a paper towel on the kitchen worktop and tested it with a fork. 'That's one of the better-tasting ones. You'll like it,' he declared. This badger was so good, he thought, because he had put it straight into the freezer. 'You can't have a badger that's been hanging around for days. Badgers have got to be really fresh, not like deer.'

Loitering in the kitchen as Jonathan cooked, I picked up a fork and stabbed the well-fried hind. It was chewy and livery and extremely pungent but not immediately repulsive, and not half as memorable as other weird things I had eaten, such as puffin or eel. Jonathan watched my ambivalent reaction closely. 'Remember this is one of my least favourite animals to eat,' he nodded, 'so it shows you how good the other animals are.'

Perhaps many of us are dissuaded from eating free-range wild meat unlaced with preservatives or hormones by the reactions of our peers. When passing motorists spotted Jonathan picking up roadkill they would blast their horns at him as if he was committing an atrocity. Their reaction, as he put it, was '"Oh my God, he's picking up a dead animal! Close your eyes children, there's a horrible man doing something horrible!" People are so disenchanted, they are so apart from nature these days.' And he believed our relationship with badgers was in 'a right old pickle'. *Meles meles*, he thought, was a 'scapegoat' for bovine TB.

Jonathan was now chopping onions, leeks and carrots and throwing them into a separate frying pan. Usually he would add wild mushrooms but there weren't many around at the moment. The kitchen was thick with smoke from the frying badger.

The *Mail*'s portrayal of him having lived solely on roadkill all his life was an exaggeration, but he rarely bought meat. 'Occasionally if there's nothing to eat I might buy a free-range chicken, but generally I don't because I eat so many pheasants,' he mused as he chopped. 'Yesterday I had a buzzard that somebody had mowed over. You take its breast out, and that's a really nice-tasting bird.' He leaned into the fridge and pulled out the remainder of yesterday's buzzard and, without asking me if I fancied some bird-of-prey appetiser, popped a bit in the frying pan with the badger for me to try. After a couple of minutes, the fried buzzard was forked out onto the worktop. It was drier and chewier than the badger and tasted not unlike the desiccated brown meat on the underside of a very overcooked pheasant. 'It was an old bird, actually,' revealed Jonathan after I delivered my verdict.

With the badger seriously frazzled, Jonathan added bean sprouts, and generous lashings of Asda's hoisin sauce – perhaps in deference to my untutored palate – and dolloped his badger stir-fry onto two plates, which we took through to the living room. He sat at the table under the window in front of his laptop; I perched on the sofa and tried not to catch the glazed eye of a stuffed albino badger, peering from a green plastic sack at my feet.

The first mouthful was fine and then, suddenly, the mound of badger on the plate before me looked rather daunting. The meat was very dense, chewy and unpleasantly strong. Like the sensation that hits you on a fairground ride when you judder from enjoyment to queasiness in seconds, I realised I was in danger of vomiting. I politely kept on chewing and tried to take my mind off my plate. Helpfully, we talked about Jonathan's life with badgers.

On paper, Jonathan might seem spooky, with his bachelor flat full of dead animals, warthog tusks, plus teeth from a tiger, a leopard and a sperm whale alongside the binoculars and sharp knives that are the tools of the naturalist-taxidermist's trade. In such exotic surroundings, however, he rather shrank from view, an amiable, slightly mournful middle-aged man with a goatee and greying fair hair. I guessed he was shy, because he did not always meet my eye, but he was gently unrepentant about who he was. As well as his big-cat tracking and taxidermy, he ran the zoological section of the Bournemouth Natural Science Society, a private museum. He had not had the opportunity to go to university and was self-taught in everything he did.

Some of his earliest memories were of the natural world. He remembered catching lizards in the garden when he was two and lived with a foster family on the edge of Bournemouth. Like Grandma, Jonathan was a twin and not particularly close to his sibling; like her, he sought out the natural world on his own. 'Two is a crowd, usually,' he said. 'If you're into wildlife you know how to blend in and be quiet.'

When he was thirteen and living on the mid-Dorset downs, he discovered badgers. 'I didn't follow football or go to the pub. I was out in the hills every night as much as possible, watching badgers and deer and owls. It was a wonderful opening up of a new way of life,' he said. 'I loved being with nature. It was my release. It was an escape.' He mapped their setts, and, after two years of watching, despite some unpleasant encounters with badger baiters, the animals would approach him and attempt to remove his shoes. 'They considered me as part of the woodland. They weren't afraid. It honoured me. I had been in care all my life. I had far more respect for animals than human beings. We're meant to be the most intelligent of species and yet we are so dishonest and nasty.'

His childhood with badgers, and his protective feelings towards them, made me think of Chris Ferris, a small, strange woman whose insomnia due to a bad back drove her to watch badgers all night. She wrote a brilliant, almost hallucinogenic account of her run-ins with badger baiters in the 1980s called *The Darkness Is Light Enough*. Jonathan adored that book and started to write one of his own at school, 'but I dropped that because I'm no good at keeping to things'. He laughed, a surprisingly deep, genial sort of chuckle that seemed to belong to a different kind of man.

Quite recently, Jonathan had joined a trip organised by the Centre for Fortean Zoology to northern India to hunt for the Yeti. They found strange footprints along the dried-up beds of streams, overturned rocks, and talked to locals who spoke of seeing giant gorilla-type animals that stood up to eight feet tall on two legs. Jonathan believed the Yeti could be *Gigantopithecus*, our giant hominid ancestor.

I was glad of the hoisin sauce. And the vegetables. Finishing my plate was proving to be a slow process. I had eaten succulent, tender lamb cooked on a Turkish grill the previous night. Badger was definitely not as good. It was not greasy, and nothing like pork, but it reeked as strongly as rancid venison or a cheap cut from a billy goat of pensionable age.

I tried to focus not on the stinking badger flesh but on what

Jonathan was saying. The Centre for Fortean Zoology, he conceded, was 'fringe science. Mainstream science won't believe it – they say it's a bunch of idiots who believe in UFOs and ghosts. Well, we do. There are plenty of people who are open-minded enough to know all these weird things exist.' The Centre hunted for monsters. I didn't realise, until Jonathan told me, that the original definition of 'monster' was 'an animal unknown to science'. He was full of learning, and wished he had gone to university; but, on the other hand, he was scornful of academics who dogmatically believed what they were taught and did not test their assumptions with direct experience. Jonathan spent more hours in the field than most academic naturalists, but then he also granted himself more leeway to make untested claims than most academics. He reminded me of Ray Mears, the countryman turned TV personality whom I once interviewed; both seemed gentle, solitary, and slightly scarred by their experience of humanity.

When he told me how he watched motorists deliberately run over injured animals on the roads, I wondered if his vision of humanity was as bleak as my grandma's. Perhaps drivers were simply unseeing, or believed they were putting a suffering animal out of its misery. Jonathan did not give them the benefit of the doubt. 'They just don't care. They don't see animals as like humans. Animals have their lovers. There is no thought for the animal that has been married for twenty years. If one of those animals dies, it's so sad for the other animal. People don't realise that animals are more loyal to each other than humans are – they don't beat their kids up, they do all they can to make sure they have a good start in life. There are sick people in this world now.'

My plate was clean at last. Jonathan was impressed. 'Most people don't finish their plate. They take a bite and then they say, "I can't eat all that."'

The spare cuts of badger hind were put in the fridge, but he had no plans to make gravy from the feet or soup from the tail of what remained of his badger in the bath. 'Tonight I'll take him out and dump him before he starts smelling my flat out,' he said. I hoped no one blasted their horns at him – although it would unnerve me if I saw a man swinging a large dead animal from car boot to hedge.

When I left Jonathan's house and headed to the railway station, a wintry day was already drawing in. I had eaten badger but I did not feel that I had crossed a Rubicon in terms of my relationship with *Meles meles*, or with Badgerland, or with society, as Angus Wilson's novel would have it. I did not feel I had betrayed the species, as Henry Williamson felt when he was blooded after watching a badger dig. In fact, I felt free, as if I had escaped everyday life and entered another realm that was beyond the comprehension of most sensible people. Who else at this moment was travelling on a train from Bournemouth smelling of fried badger?

The pleasure I took in this deviant behaviour, and the more rational sense that if we were to live in harmony with nature and eliminate waste then we should all consume roadkill, was not enough, however, to overcome one unsurmountable obstacle: badger tasted bad. I realised that our reluctance to eat it was not because of a taboo against devouring well-loved wild creatures or the anthropomorphic animal heroes of children's books. Badger has always been a meal of last resort and in medieval times the rich did not go near it. As the Master of Game to Henry IV wrote of 'the grey': 'His flesh is not to ete nor that of the fox.'

Every so often, as my head lolled towards the shoulder of my jacket with the rocking of the train, I could detect a lingering whiff of Jonathan McGowan's kitchen, fuggy with the fumes of fried badger. It took all my powers of concentration to stave off a desire to vomit, which was amplified by every badger-flavoured burp that escaped my mouth.

14

Bella

The Cornish countryside had been scoured by the changing seasons, stripped of its summer greenery and of tourists. On a clear cold day, the sky was a brilliant blue and the grey roads were salt-flecked and emptied of all but speeding locals. Badgerland was convulsed by debates over the cull but I had another priority: Judy Salisbury's badgers had abandoned their home sett and I wanted to find them.

That autumn, Judy rang me out of the blue and told me her arthritis had taken a turn for the worse and she was struggling to feed her badgers. They had also been flooded out of their sett. One Wednesday night in November, she said, it had rained harder than she had ever known it. The following evening no badgers came for their nightly feed on her patio. When the rain stopped, still no badgers. No Willow. No Salt, Pepper, Mustard or Vinegar. Judy was convinced they had drowned. After the badgers disappeared, she put food out for them every night, her usual banquet minus the sausages (cooking them would have been too wasteful). After three weeks, she gave up, although she still made up a small box of sandwiches and placed them by the *Radio Times* and the magnifying glass on the coffee table by her armchair. Just in case.

One night shortly before my arrival, Judy was watching *EastEnders* (an incongruous passion) when she twitched back the curtain and caught sight of two badgers trotting off her patio. She grabbed a sandwich and hurried as fast as she could to the door. 'I just said, "Come along"' – she slipped into her special high-pitched badger quaver – 'and they turned. That was the happiest day I could remember for a long time because I knew they were mine.' They came back, ate their sandwiches and, since then, some of her old family had visited for food, albeit irregularly and often late at night. She was convinced they were arriving from further afield than before. And she had still not seen the matriarch, her favourite, Willow.

When I walked down the pathway to her front door, Judy was standing there like a sentry. Moving stiffly, she showed me through to the spare room with its not unpleasant mossy scent, the dampness of a room rarely used. I did not realise how rarely until she revealed I had been the last person to stay there, six months before.

I decided that the best way of finding out where the badgers had retreated to would be to trace their habitual route to Judy's garden. In the last of the afternoon light, I set out on the public footpath across the field behind her house. The salty edge of the creek had iced up. The horses on the hill on the other side of the estuary were swaddled in green and red blankets. In the still, frosty air, a curlew delivered its delicious gurgling warble, a bubble of mud rendered into birdsong.

The paddock was pitted with a venerable badger sett. A council sign warned walkers of the hidden holes. 'Please take care until we can permanently resolve the situation,' it said, which sounded ominous. Beginning inside the blackthorn thicket that bordered Judy's garden, the sett had moved into the field like ribbon development. Before the flood, the residents must have scratched themselves with joy at such an easy life, an as-much-as-you-can-eat free buffet laid out on their doorstep every evening.

Judy was too frail to walk onto the field these days and I had never tried tracking a badger before, but it was surprisingly easy. A broad badger highway emerged from the blackthorn and marched on across the meadow, distinct but wobbly, as if made by fairies. It dived straight through a thick hedge, and headed directly across freshly planted winter wheat – proof that it was regularly, and recently, used. At the far side, the badgers took the easiest option, straight under a steel gate, and crossed another field before carving a deep furrow into the bank below a hedge so they could cross a sunken lane. The animals then slipped past the caravans in a small park set in a disused blue-slate quarry.

Here I lost the badger trail, but Judy had been told of an active sett not far from the footpath by Little Petherick Creek. I followed the path along the high-tide line. The salt marsh was a frosty silver-green, turning pink where the setting sun brushed sand and water. Four oyster-catchers strutted across ice-stiffened mud. Lichen hung raggedly from scrawny bushes and trees. The grandest ash was almost bonsai in scale, buffeted by the south-westerlies.

Cawing rooks took to the sky. A mile from Judy's house was a meadow full of squat thistles; mottled blue with frost, they resembled jellyfish dotted in a green sea. Beyond was a plantation of cherry trees, perhaps twenty years old, enclosed in a deer-proof fence. There, under the wrinkled old blackthorn hedge surrounding it all, was an enormous sett. I imagined the tunnels reaching like tendrils under ground, giving bloom to messy holes in the plantation. At the top of the field, a stone wall had collapsed and I guessed the badgers took this short cut to begin their journey to Judy's nightly banquet.

I was moderately proud of my tracking – it gave me a fleeting sense of being a badger on my nightly ritual: rising at dusk and following the pathway made by my trusted friends and relations to a marvellous source of food. I retraced the badger highway back to Judy's with what I hoped would be cheering news for her, although I could not really tell if it was, as she was now preoccupied with her evening routine. She cooked me tea and tried to stop me washing up before pulling on her coat and blue bobble hat. At 7.25 p.m., she slid open the patio doors and began throwing food onto the patio with her green plastic children's spade. I offered to help. 'It's better that I do it on my own,' she said. The room was completely dark, except for a feeble orange glow from the wood-effect electric fire, which provided no heat whatsoever.

Judy was troubled by the continuing absence of Willow. The apparently leaderless group now usually arrived in the middle of the night, when Judy wasn't watching. The most she had seen together since the flood was four, a far cry from the ten of midsummer.

When Judy finished dishing up, we closed the curtains so as not to disturb any tentative approaches and turned on several dim lamps. When *EastEnders* finished, she checked behind the curtain again. None of the food had been taken. In the chill of the frost, the badger-way across her lawn stood out like a big grey stripe.

Over the estuary, we could just make out the obelisk on the horizon. I never asked Judy about this on my first visit, but now she explained it was one of Queen Victoria's memorials to Albert. In a hedge bordering the estuary, she revealed, her husband Robin had planted a stand of pampas grass twenty-five years ago. 'When I look at it, it's a memorial to Robin,' she said. 'I like pampas grass. I'm fascinated by the movement.'

Judy went to bed and I stayed up for a while, twitching the curtains now and then, but no badgers came. When she woke me at 7 a.m. with a colossal thump on my door and the forbidding phrase, 'Time is getting on', all the food had gone. In the misty morning light, I saw for the first time the frosted tufts of pampas grass, protruding from the hedge like triffids.

Over the past months, I had seen wild badgers at closer quarters than I imagined possible, and these communions with them were thrilling and satisfying. Their lives, and those of the human residents of Badgerland, felt like they unfolded in another dimension, intriguing and strange. But I did not fully realise what I still lacked in my life until I bumped to the bottom of the track that sank towards Bottengoms, Ronald Blythe's home on the border of Suffolk and Essex.

It was a limpid autumn evening and a big ash tree that loomed over Ronald's garden was instantly recognisable from his writing. 'All the pruning is done by gales,' he wrote in *At the Yeoman's House*. 'Now and then pairs of magpies dive snowily from a particular bough like Olympic youths, white and perfect, to seize crusts.' When I knocked on the door, an orange lamp already cast a warm glow across the old kitchen, where Ronald was pottering about at the sink.

He was an inspiring person, approaching his ninetieth birthday and still writing books, as well as a weekly column for the *Church Times*, from his old farmhouse on the edge of the Stour Valley. He first walked down its track when he was twenty-three, a local reference librarian befriended by Christine Nash, wife of the painter John, who had bought the property four years earlier. An aspiring writer, Ronald was drawn into a glamorous, bohemian, artistic set comprising the Nashes; the painter Cedric Morris, who taught Lucian Freud; Benjamin Britten, for whom Ronald compiled an anthology about Aldeburgh; and even E. M. Forster. After the success of *Akenfield*, Ronald's portrait of a Suffolk village written through the first-person accounts of its residents, which became an instant classic at the end of the 1960s, he nursed his friend John Nash in his declining years. Eventually, Ronald came to live in Bottengoms himself, with his cat and his books for company.

From the way he spoke and thought to his old Suffolk surname, derived from the River Blyth, Ronald was perhaps more deeply rooted in the countryside where he was born than any other living writer. But he was not some kind of rural hermit: he was widely travelled, fascinated by the world beyond, and appeared to remember everything he read ('I find it not difficult to remember things,' he said in his understated way). He was not simply an East Anglian writer, either, although the decades he spent in its frugal landscape, shaped by the wind and the sea, must have seeped into his sentences. As he gently pointed out, he was, simply, a writer who lived in the countryside. Frustratingly, I could not call myself that, although I was on the brink, finally, of leaving London, where I had lived for most of the last fifteen years. Lisa, my girlfriend, was expecting twin girls, a glorious surprise, and we calculated it would be easier to cope if we moved back to Norfolk, within the orbit of both our mothers. At last, I had an escape route out of the big city, and a chance, perhaps, to forge a more permanent relationship with a piece of countryside, and a sett of badgers. My current status was of a tourist in Badgerlands. I did not belong there, or anywhere.

I had returned to Ronald's door after I interviewed him for the *Guardian* and he casually revealed in passing that he had a sett in his back garden. He had not studied his badgers but knew they were there, in his overgrown two acres. They were a habitual part of his easy intimacy with the countryside around him, a rich, deep connection I could only envy. 'Sometimes you come down the track and see them cross your path in the evening,' he said. 'They often squabble, don't they?' He had a humble habit of turning observations into tentative questions.

We sat in Ronald's kitchen and talked about darkness as the sky through the small windows of Bottengoms turned inky blue. 'Like

Robert Frost, I can say that "I have been one acquainted with the night",' wrote Ronald in *Word from Wormingford*. 'I enjoy a ringing step on a night road and making out the broken-backed roofs of ancient farms, and searching ploughed hills which at such a moment could be going up and up until they become alps.' Although he now required glasses to watch television, Ronald was blessed with excellent night vision, and roamed the countryside in varying degrees of darkness. He told me how Coleridge and Wordsworth and his sister used to go walking together at night when they were in their twenties and staying in Somerset. Coleridge was doing nothing more dastardly than writing *The Rime of the Ancient Mariner*, but Britain was at war with France and local villagers feared these night walkers were spies. 'Everything changes at night. The trees change. Even places you know backwards take on another life at night. They become mysterious,' said Ronald. 'I don't find it fearful but there's a history of people finding it fearful.'

This historic fear of the dark has made us afraid of badgers, and it may have made us more brutal towards them. For all that was elegiac in *Akenfield*'s depiction of a vanishing time, Ronald was not sentimental about the old ways. He showed the 'glory and bitterness' of country life, as he put it: its poverty, ill-health, incest, and the exploitation of labourers by their masters. But we shared a melancholy sense that most of us, even village dwellers, have lost a profound understanding of the land. 'There is often very little true life,' he said of contemporary rural existence. 'In a remote village, people lead the lives they might lead in Birmingham or anywhere, with the same TV and fitted carpets and supermarkets. I am really quite disturbed that almost nobody in the villages now knows anything whatsoever about the land. And they don't look at the fields.'

Nevertheless, the countryside itself, judged Ronald, was far healthier than when he wrote *Akenfield*. Gone was the uncompromising industrial farming of the 1960s and 70s, which ripped out hedges and sprayed dangerous pesticides; gone was the grinding physical labour, when people literally worked themselves to death; and gone was the cruelty of the rural landscape – for all its inhabitants. Gone too was the old barbarity towards badgers. Baiting 'always happened here and it was a very disgusting thing,' he said. 'When I was a boy, the young men would walk around the fields with a gun shooting everything they could see. Terrible.' Not long after he inherited Bottengoms from John Nash, a smartly dressed woman of the hunt turned up. 'We're going to draw here,' she announced presumptuously, and turned savage when Ronald quietly told her that no, he would not permit the hunt to rummage around in his badger setts looking for foxes.

A pheasant's pre-roosting shriek alerted me to the urgent need to get into position for the badgers. The sun had already set and a frost would soon fall. Ronald announced he would stay inside and 'have a little badger hunt upstairs', looking for badgers in the treasure trove of all the books he knew, and all the books he remembered.

The rapidly cooling air pinched my cheeks. The breeze in the trees sounded metallic now, as if every brittle leaf was made of silver paper. The boundary between Ronald's garden and his two-acre wood was blurry. Barely thirty yards from the great tiled roof of his farmhouse, which turned its mossy shoulder on the trees behind, were the first badger holes, beneath a tangle of ash, oak and horse chestnut.

The previous weekend I had met a very interesting zoologist turned sculptor of precious metals. When he asked me about badgers and bovine TB, I told him I was trying to sit on the fence. 'Good. Fences should be sat on until they are completely squashed down,' he said rather fiercely. It was one of those unnerving moments when I had no idea in which direction his thoughts were travelling. 'We are always trying to put things in boxes. Divide things up. Fence things in. There are no fences in the natural world. The clearest boundary you get is a river, or the coastline. And that is not really a boundary at all. It is not *hard*. Things pass across it.' The zoologist's philosophy reminded me of one of the practical preoccupations of the badger cull: Defra's belief that it would be more effective if the culling area was surrounded by so-called 'hard' boundaries, such as rivers or motorways, which would stop any badgers that escaped death from fleeing and spreading disease. This faith in hard boundaries sounded like a fantasy. So did my sitting on the fence.

In the natural world, however, boundaries are places of transition and beauty, especially when they involve water. My favourite kind of non-watery boundary is the collision of a bulging woodland with the curve of a field. This boundary is the badger's favourite as well, because a contented life indeed is a woodland sett by a pasture rich in earthworms. Ronald's wood met the ploughed field beyond in a particularly satisfying embrace, and there, on this gently curving line of beauty, I found a field maple with a fork into which I could climb. It gave me a panoramic view of the murky interior of the wood and a badger track wiggling across the woodland floor.

Wedged here, I waited. Geese honkingly complained about this and that, three fields away. A distant cow lowed. A rustle, and I was alert. A tubby chestnut-brown muntjac with pointy antlers hove into view. This Mr Tumnus of the woods picked his way through the crackling undergrowth on his evening rounds, passing within a few metres of me. I felt a stab of satisfaction: I hadn't been rumbled on my perch. Another rustle, and a brown mouse skipped across the leaves at a surprising pace. How could an owl catch such a little scooter?

The great acoustic blanket of midsummer, those dampening green leaves, had been lifted; grounded leaves also crackled and popped, amplifying every footfall. Despite this, I didn't hear it coming. Badger! A pale-grey shape beneath me, barely ten yards away. I held my breath. I sensed it was turned towards me, swirling and twirling the air with its nose, inspecting either the bulky silhouette I made against the sky or the suspicious smell I must have discarded when I trampled my way to the tree.

Eventually, it had its fill of smelling me and moved off again, a pale, oblong shape in the darkness gliding over the leaves like K-9, the robotic dog from the 1970s *Doctor Who*s. There was not so much as a crunch from the woodland floor. A second badger was noisier, and boasted more distinct black-and-white markings. Two badgers and me, in a wood on the border of Suffolk and Essex. I didn't feel like I was intruding on their space; I belonged here too. For the first time, I was completely at home watching badgers. The sweep of the fields and the sparse hedges dotted with stag-headed oaks were recognisably East Anglian. So was the kark of the fat, careless pheasants. They might not be native birds but they were part of the landscape I had grown up in and they made me feel irrationally joyous. I imagined they were calling me back to my heartland.

At long last I had seen some badgers at a sett by myself, without the aid of experts or peanuts. For half an hour they foraged in the wood below. They would appear in my field of vision quite suddenly, and then, frustratingly, melt into the shadows where my eyesight could not follow. I could hear the badgers more clearly than I could see them, but sensory data was scant: for a creature that laboured under the reputation of being a noisy blunderer, these animals were astonishingly light-footed. Perhaps one in five of their steps crackled a leaf. When I lowered myself from the tree, I cracked three twigs with my first stride.

The badgers and I shared this twilight. Of course this pair had not welcomed me into their world and were only beetling around below because their nasal risk-assessment had concluded that my smell did not pose a clear and present danger. But I still felt an accepted part of an hour, one evening in East Anglia. In a beautiful passage in his memoir about growing up on a farm in Herefordshire, Horatio Clare wrote of the joy of watching a badger family. 'We had known the edge of an evening wood as a badger knows it. For that twilight time we had slipped the separation between us and the world. We were of the mountain, and of the wood, and it was as though the animals, the wild creatures, had allowed it. It was bewitching.'

I stumbled noisily into the field and the stars exploded overhead. Three planes were visible, stacking for Stansted airport. A fox barked, hoarsely, six times; next up sounded the rich, watery relief of a horse in the field opposite taking a long pee. I lay on my side, on the wavy line between forest and field, and listened to the badgers' foraging music inside the wood. They were feeding, meandering through their home territory with an occasional leafy scuffle, interspersed with more exotic notes: a yawning, a scraping of claws on bark, and a mischievous noise that sounded like skittles being knocked over. Then came a distinctively canine crunch, like a dog with a bone. Was it a badger with a particularly large beetle?

The badgers never abandoned the security of the wood for open field but I did not mind. I looked up and the vapour trail of a plane descending towards the airport was gently illuminated by the over-the-horizon glow of the lights of Colchester. As ghostly as a cobweb, this nocturnal trail appeared to suspend the plane from the sky, as if it was a toy in a child's bedroom.

Suddenly I was shaken by a deep chill. I staggered up and walked around the field, skirting Bottengoms, before returning to Ronald Blythe's house. In the darkness of his garden I snapped on my torch and there, on his front lawn, was a badger, turning towards me, its small eyes glittering in the light. A crash, loud this time, and it was gone.

After his hunt upstairs, Ronald recommended I read W. H. Hudson's writing on badgers. I told him how delighted I was to have seen badgers in his wood, but I could not quite convey how I had been touched by the evening. I felt like I had come home and I saw that home need not be where you began life, and need not be unchanging either, as this night, my first experience of East Anglia to include badgers, showed. But truly being alive to the possibilities of home required a deep commitment to being still and, perhaps, alone in one place. This I lacked. As Ronald said to me, 'If you go for walks with a friend in the countryside that is a lovely experience. But if you live as I live in the middle of nowhere by myself, that's another experience. There's nothing mystical about it but it makes me dream. If you're in this house, surrounded by fields every day, something happens to you. I don't know what it is.'

The approach of the badger cull seemed to herald a decline in Judy Salisbury's health. 'It will kill me,' she declared of the cull when I phoned her one evening. The badgers in the proposed cull zones in Gloucestershire and Somerset may have been under threat but in Cornwall, Judy's flooded-out badgers were safer than ever, having returned to the sett close to her garden. Eight animals were getting extremely well fed each night. Judy, though, was less comfortable, for the oil tank providing the tenuous heat emitted by her radiators had burst six weeks before and had not been fixed. Her boiler had packed up too, leaving her without hot water as well as heating.

It sounded scandalous that a woman in her eighties, living alone in a chilly house, had been left without central heating for six weeks of a cold, wet autumn. I hoped to watch her badgers again but I also wanted to knock some heating engineers' heads together and so I invited myself round. 'You will have to take me as you find me,' said Judy, rather forlornly. 'I am very much more disabled now.'

Daylight was disappearing when I slipped down the tiny lane to Judy's, a sense of calm enveloping me like cool air sinking at the end of the day. It was a kind of magic conjured up by the innate silence of this small valley; the tide on the marshes; the organic, slightly damp scent of Judy's home that reminded me of my grandparents' house; and the way her road drifted downhill towards the sea, sinking further into the earth as it did so.

'I'd given up on you,' said Judy, boiling the kettle for tea. By her front gate was a rough crater where her oil tank once stood. An air of dereliction suggested that no workmen had been sweating here for some days. Things were not always what they seemed with Judy, however. It turned out that the engineers had at least installed two temporary oil tanks to provide her with central heating. The boiler had also been patched up. Assisted by some portable radiators, the house was warmer than I had ever known it, which was not very warm. I suspected she had only turned on the heaters for my benefit.

It was not easy to help such a resourceful person. I had stopped at a supermarket and bought her a new electric heater but she said it was not suitable. I brought pasta and vegetables to make her some tea. 'I never have supper,' she declared, which was news to me. I tried to make the tea but she insisted on dunking her tea bag in the hot water herself. I offered to hassle her errant heating engineers but she insisted she had already telephoned them and given them 'a rocket'. There was nothing to do but retreat to Judy's crepuscular front room and wait by the French windows for the badgers.

It felt like we were at the cinema. On my lap I balanced a bowl of crisps, which I tried not to crunch. Judy sank into an armchair by her coffee table on which rested a magnifying glass and the souvenir programme from St Issy & Little Petherick's Diamond Jubilee celebration – featuring Fairground Organ Music, Dave's Disco and Many Other Attractions. I glanced across at her silhouette. As she had warned, she was even more disabled, and both her legs were stiff and stretched out straight in front of her. In the half-light, she looked like a thin, graceful doll that would not bend at the knees. When she had a coughing fit, she hauled herself out of her chair and shuffled from the room. 'I don't like coughing in front of people,' she explained.

Somehow, despite not eating and moving with a fearful tremulousness, she continued to assemble her nightly badger feast: a washing-up bowl of soaked dog biscuits, stale buns, banana, apple and broken-up pork pie, 'the little cheap ones' from Tesco, as well as four two-litre tubs of sandwiches, a four-litre tub of peanuts, a container of sausages and a tub of grapes. 'They love them. They take a single one in their mouth and put their head back and go, "glug, blub, gurg",' said Judy, imitating a badger in ecstasy.

Around each container was a thick yellow rubber band.

'Would you like some of these elastic bands?' she asked.

Badgerland had the knack of making me feel exceedingly stupid. I did not know what to say.

'They are Marigolds and they never break,' she explained, chuckling. 'I use them all the time.' And so they were: washing-up gloves, cut into bands.

High tide brought twenty-six swans gliding across the navy waters of the estuary below. One of the swans was black. An omen? Judy suspected it was an escapee from a wildlife park. The texture of the land was rubbed out by the twilight and an hour turned into two hours. A weasel was briefly illuminated as it bounced across the lawn; bats darted through the brightness spooling from the patio light; a mouse wheeled into the shadows on the patio. The tang of oil from a far-distant boat drifted on cold air through the open patio doors. We spoke of a lesser spotted woodpecker Judy had found dead on her lawn, the loss of barns and barn owls, and Judy's trick when salespeople phoned up: she kept a toy whistle from a cracker by the phone and every time someone introduced themselves she would blow the whistle as loudly as she could. They soon hung up. Most of all, we talked about the cull.

'I am convinced it's bad husbandry,' said Judy, of bovine TB in cattle. 'You get cows shut up, cheek by jowl, eating God knows what. They are born, they live without seeing a blade of grass. It's an unnatural way of living.' I wondered aloud whether, if bovine TB did not exist, we would be debating another reason for killing badgers. Why did we display such persistent ill-feeling towards them? 'There's a part of humanity which just wants to kill. It doesn't really matter what it is,' said Judy. 'We have been given life and we have no business to take it.'

Shortly after 7 p.m., a mask materialised in the shrubbery at the bottom of the garden. A perfect constellation of black-and-white triangles held there, motionless, for a minute.

'Come along. Supper time,' called Judy.

The mask moved, and then withdrew again, as if it had been a figment of my imagination. Something was making the badgers nervous tonight.

When the first one emerged from the hedge, it moved with an unusually pronounced bumble. 'She's limping,' said Judy sharply, noticing it straight away. 'It's Bella, our oldest badger. I think she's deaf.' In the stillness, the rat-tat-tat of peanuts thrown onto the patio was almost as loud as gravel against a window pane, but the animal did not respond. Judy dispatched her feast with her usual aplomb although she complained she couldn't hurl the sandwiches very far. 'This wretched soft bread, it's difficult to throw it's so light,' she said.

Old Bella limped off after twenty minutes of foraging and the white squares of Judy's uneaten sandwiches stood out on the lawn like the swans on the dark water.

The grandfather clock struck a strangulated eight. Judy had muffled its strike because 'the badgers didn't like it at all'. Every time I visited her, I learned at least a dozen new small, strange things.

From beneath the elder came a whickering and then a decisive quack: the call of a mother badger sounding like a moorhen, just as the books said. More badgers were gathering. A second and a third emerged. A fourth flicked up its nose, looking dismissive but actually assessing the notes of danger and promise in the still night air.

'Would you like a sausage?'

Judy shuffled to stand by the supper trolley. 'We may be in business now. Come along, chaps.'

A few flings of her plastic spade, and the patio was pebble-dashed in dog mix. There were four on the patio steps, and then eight, eddying this way and that, as if they were connected, a toy train of animals moving to Judy's beat as she rhythmically distributed food from the washing-up bowl. The wet sound of the badgers eating was as relentless as the trickle of the stream into the estuary. I lay down, half leaning out of Judy's doorway, and breathed in the tang of dog biscuit on the night air and, churned into this, the warm, horsey smell of living, breathing wet badger.

This time, I knew not to ask whether I could serve dinner; that priestly function belonged to Judy alone. Her garden was transformed in the dusk. Pigeons and cats had absented themselves; badgers ruled this land now. They appeared to move slowly but looking more closely I saw that their jaws were moving as relentlessly as a mouse's heartbeat, efficiently snaffling every morsel.

Their absorption, their utter comfort in the fine dining they found nightly in Judy Salisbury's garden, reminded me of the scientists' comment that foraging badgers were like shoppers in a supermarket: when there was so much food, it was a waste of energy to fence off individual territories. Instead, family groups ate together, tolerantly stepping around shoppers from other social groups. My admiration for this feeding fraternity was suddenly halted by a thought so vivid it was as if I had seen it. At some point soon, this exact scene would be played out at apparently generous and benign badger supermarkets built conveniently close to setts in Gloucestershire and Somerset. Having taken all the usual precautions, a dozen badgers would be browsing the aisles together and then a barrage of shots would ring out. Most, hopefully all, would perceive a flash of light, a punch in the guts like nothing they had ever experienced, before darkness descended for ever. A few unlucky animals would stagger off, nursing horrendous injuries, if they were not wiped out by the second, or third, volley of shots.

'What a show,' sighed Judy. 'That was worth waiting for.'

I didn't want to tell her what I had just imagined. The cull around the corner weighed heavily on Badgerland.

15

Bessie and Baz

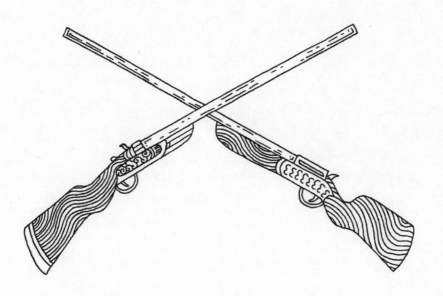

Malcolm Tucker, a potty-mouthed fictional spin doctor, first voiced a memorable phrase to describe the multiple failures of the hapless government in *The Thick of It*: it was, he said, an omnishambles. Like so many features of the satirical comedy, the word was swiftly commandeered in real life to describe the confusions of the coalition administration. When the government became entangled in a mighty controversy over badgers in 2012, the phrase morphed once again in social media. On Twitter, the badger chaos became an #omnivoreshambles.

The omnivoreshambles had everything that delights the British media: moral outrage, politicians under fire and panicky policy U-turns. Writers of fiction would struggle to invent such mayhem: one long-haired rock star fighting for the rights of badgers on breakfast telly; two shadowy companies created specifically to orchestrate the cull; a hundred vuvuzelas, bought on eBay by animal rights activists to scare the badgers from the feeding stations built to lure them to their deaths. Most deliciously, above the attractively variegated cast of bungling politicians, pitch fork-brandishing farmers and animal rights nutters, loomed the photogenic hero: the noble, innocent, tragically flawed black-and-white.

Forty years of turning and U-turning over badgers and bovine TB were perfectly distilled in a few months of 2012. This was the apotheosis of a stalemate that has been classified in learned academic journals as 'policy failure'. Generations of farmers have failed to stop their cows contracting the disease. Generations of civil servants have failed to steer politicians on a steady course of action. Generations of scientists have failed to establish precisely how the disease is transmitted between cow and badger. The only constant has been the badger, and the unwavering affection in which it is held by most of the public.

A cull of badgers was promised by the Conservative Party in its manifesto before the 2010 election. The coalition agreement with the Liberal Democrats reiterated this pledge to introduce 'a carefully managed and science-led policy of badger control in areas with high and persistent levels of bovine tuberculosis'. The following summer, with most people preoccupied by the deepest economic downturn since the nineteenth century, the Environment Secretary Caroline Spelman confirmed a badger cull would go ahead in England. The mood in Defra's stuffy building in Whitehall reminded me of the government in the months before the Iraq War. There was no alternative, and everyone from the chief vet to the department's chief scientist was toeing the party line, and making sure the dossiers of data fitted their decision. Confusingly, in Wales, where the devolved government had long planned to cull badgers in Pembrokeshire, there was another abrupt U-turn when the Labour administration elected in 2011 rebuffed the advice of its chief scientific adviser and cancelled the cull, announcing that TB in cattle would be tackled across Wales by a five-year badger vaccination programme instead.

Government grinds along slowly, and in England there was another consultation in late 2011 before the announcement of a 'pilot' cull in two bovine TB hotspots: west Gloucestershire and west Somerset. If this cull proved successful, it would be expanded to up to ten areas in England in subsequent years, a rolling programme of killing badgers that would have no end, until bovine TB in cattle was vanquished.

What began as a flat and rather indistinct-sounding term, *a cull*, steadily acquired the kind of detail that horrified many ordinary people. It was called a 'farmer-led' cull because farmers had to make a financial contribution, but in reality it was heavily funded by Defra. The way that landowners' and cullers' anonymity would be protected by the creation of special culling companies seemed sinister. Worse, no one except affected landowners would be told the precise location of the culling areas, which had to be at least 150 square kilometres in size and possess clear physical boundaries, such as rivers or motorways, which would reduce the likelihood of fleeing badgers spreading disease. Each cull would go ahead only if landowners who owned 70 per cent of the land in the culling zone gave their financial support, and at least 70 per cent (but no more than 95 per cent) of all badgers in the area, healthy and diseased, would be killed. The culling method was 'controlled shooting': masked gunmen with night-vision goggles or lamps would fire on the wild badgers lured to special feeding stations from a distance of around seventy metres for a period of six weeks. Every kill would be recorded on the cull company's database and carcasses would be placed in two bags. Dead badgers would be either incinerated or rendered.

The cull would start after the Olympics to ensure, it was revealed rather ominously, there were enough police officers to maintain public order. The Games flashed by, a holiday from the ordinary world, a rare moment when most people in Britain felt good about their country, and said so. But as fireworks sailed high over the Olympic Stadium in Stratford and celebrities adorned the closing ceremony, Brian May appeared on stage, playing guitar alongside Jessie J. Sewn on his jacket sleeves were two badges: a fox and a badger. The rock star had smuggled a pointed political statement into an anodyne, avowedly apolitical celebration. This gesture, in its own small way, was a declaration of war. The badger became headline news, and all hell broke loose.

Would culling work? Both proponents and opponents of the cull had cherry-picked the science that supported their arguments, so I returned to the Cotswolds to meet the man dubbed by one local 'the Pol Pot of badgers' for his leading role in the Randomised Badger Culling Trial, the epic, seven-year test of the efficacy of culling badgers that began in 1998.

Chris Cheeseman was a muscular sort of scientist. He was clean-shaven, wore a smart North Face shirt, drove a BMW and lived in a comfortable modern house on the edge of a sunlit Cotswold valley.

Cheeseman had always been interested in science in the real world and, in 1975, received a brief from the Ministry of Agriculture that would define his life's work: 'Badgers have been implicated in the occurrence of TB in cattle. Can you study their biology and learn about their involvement?' So he scouted a suitable territory for badger experimentation and set up Woodchester Park with an assistant who had worked under Hans Kruuk at Wytham. Their research station grew like the badger setts it surrounded: at its peak, it was home to forty-two scientists, thirty-six of them devoted to the problem of badgers (the rest were wrestling with ruddy ducks). Cheeseman co-authored a book with Ernest Neal, and after decades working for the government, was now retired and free to speak his mind, which he did rather impressively.

The Randomised Badger Culling Trial, led by Professor John Bourne, selected thirty separate, broadly circular 100-square-kilometre areas in the west of England and grouped them in triplets. In one zone, badgers were repeatedly, proactively culled once a year – caught in traps and cleanly shot through the back of the head; in a second, badgers were reactively culled close to outbreaks of bovine TB in cattle; the third zone was the control, in which no culling took place. Rates of bovine TB in cattle were closely monitored.

During the RBCT, Chris Cheeseman trained staff in Defra's 'wildlife unit' – the badger exterminators – and audited much of the work. During the cull, his car windscreen was smashed, he received menacing phone calls from animal rights activists, and his mild-mannered daughter punched a tormentor at school who accused her dad of killing badgers. Defra told him to check underneath his car for bombs each morning, and fitted an alarm system on his house. This, he thought, was an overreaction but the RBCT scientists got it from all sides: animal rights protesters sabotaged their work while farmers were unhappy that all other badger culling was halted during the trial.

When the first results emerged, 'everybody was absolutely stupefied. Nobody believed it,' remembered the badger biologist Tim Roper, who was visiting Woodchester at the time. The scientists had assumed that culling would reduce the disease in cattle; in fact, reactive culling was halted in 2003 because the incidence of new cases of TB in cattle – called 'breakdowns' – was 18.9 per cent higher in reactive areas compared with the no-culling zones. Something odd was happening around the proactive culling zones as well. During the seven years of the RBCT, annual proactive culling in the ten 100-square-kilometre areas caused a 23.2 per cent decrease in cattle TB breakdowns when compared with the no-culling zones. However, proactive culling caused a 24.5 per cent increase in cattle TB cases in a two-kilometre-wide ring surrounding the cull zone.

In other words, culling was making things worse.

Badger scientists had a theory for this: perturbation. When a social group was disrupted by culling, its surviving members roamed further afield than usual in search of safety, taking their disease with them; where badgers had been wiped out, other badgers moved in; more mixing between groups spread more disease. Cheeseman remembered farmers reporting in the late 1970s that culling appeared to be triggering new outbreaks of bovine TB in cattle. When he told civil servants that culling was making it worse he was ticked off for what was called 'an unhelpful distraction'. This fired him up. 'You need a certain amount of aggravation to motivate you. It made me resolve to get the evidence. It took me a long time.'

'Perturbation' made perfect sense to those who study badger behaviour. One paradox of the theory was that it offered very strong evidence that badgers were causing TB in cattle while simultaneously demolishing the case for a badger cull. 'The great irony in the RBCT is it provided by far the best evidence so far that badgers are implicated, but it seriously undermined the only way we had for dealing with the problem, which was culling,' said Roper. He sounded one note of caution. While the RBCT provided very good evidence that culling disrupted badger social groups and caused an increase in bovine TB in them, the evidence was not so strong for the third stage – transmitting it so quickly to cattle. Roper wondered how it was possible for straying badgers to infect cattle, and for those infections to be detected (by the notoriously unreliable testing regime) so rapidly. 'Since the RBCT, the perturbation hypothesis has gained in strength rather than diminished but I am troubled by the sheer speed it happened in some of the trial areas,' he said. 'Although I remain sceptical about the perturbation hypothesis in many respects, the scientific method is to believe the hypothesis until a better explanation comes up. I still think there's something wrong with the perturbation hypothesis, but it is the best explanation we've got.'

As yet, however, there is no peer-reviewed science to support

Roper's scepticism, and the scientific consensus – and evidence – over-whelmingly supports the theory of perturbation.

At first glance, the RBCT suggested that culling badgers was worse than useless at stemming the spread of TB in cattle. 'Badger culling cannot meaningfully contribute to the future control of cattle TB in Britain' was the apparently unequivocal conclusion of Bourne and his team in 2007. Lord John Krebs, whose review led to the RBCT, in 2011 summed up the benefits of culling: 'You cull intensively for at least four years, you will have a net benefit of reducing TB in cattle of 12–16 per cent. So you leave 85 per cent of the problem still there, having gone to a huge amount of trouble to kill a huge number of badgers. It doesn't seem to be an effective way of controlling the disease.'

Bourne's decisive conclusion was, however, weakened by new data as epidemiologists continued to record the incidence of TB in cattle in the trial areas after the cull stopped in 2005. Subsequent analysis by one of the trial's scientists, Professor Christl Donnelly, contradicted her own initial findings by showing that proactive culling produced appre-ciable benefits in reduced cattle infections six years after culling stopped: between one year after the final proactive cull in 2005 and August 2011, there were 28 per cent fewer herd breakdowns in pro-active cull zones than in the control areas. In other words, culling badgers might produce a fairly modest benefit but it lasted for a long time.

It is no surprise that many farmers had little faith in the RBCT. Farmers do not think scientists make good killers. Some joked that the culling was so ineffective it should be called 'the randomised badger dispersal trial'. The social scientist Gareth Enticott interviewed members of the NFU, who felt the trial was 'deliberately slanted' against farming interests. Judged on its terms of reference, the RBCT was excellent science, but terms of reference are a subjective creation. The scientists were told by ministers at the outset that the elimination of badgers from large tracts of the countryside was 'politically unacceptable' and badger welfare must be taken into account. The most effective way of culling badgers – gassing them – was banned. The approved method, trapping and shooting, was less effective. In farmers' eyes, the culling trial was set up to fail.

The cull was suspended during the foot-and-mouth outbreak of 2001, which reduced its efficacy, and was disrupted by protesters. According to an answer to a parliamentary question in 2003 (while the trial was ongoing), 57 per cent of traps had been interfered with, and a further 12 per cent had been stolen. One animal rights activist claimed that he and friends 'chopped up' thousands of traps. 'It was so easy to disrupt,' he told me. 'We stashed bolt-cutters in different locations, went around on foot finding the traps, got the cutters and chopped up the traps.'

One critique of the failure of the RBCT to efficiently kill badgers came from one of the cullers, who told a House of Commons select committee in 2006 that the cull was 'farcical' for its first four years. Paul Caruana, who worked for Defra's wildlife unit in Cornwall for twelve years, claimed that the RBCT's limited trapping (badgers were only trapped over a maximum twelve-day period per year) had little effect and that reactive culling was 'prematurely ended'. An increase in cattle TB was also observed outside culling zones and so, Caruana believed, the increases recorded around the zones may have been caused by factors other than perturbation. 'The trial has far too many flaws in it to be trusted to produce meaningful evidence,' he said. 'I know what happened on the ground – the scientists only have the results we provided them with to work with. I know that those results could and should have been much better and more useful than they currently are.'

A Defra statement in 2005 (before the trial had finished) said the culling efficacy of the RBCT was between 20 and 60 per cent, but Chris Cheeseman was bullish about the success of the RBCT. 'I don't think it's realistic to say it could've been improved upon,' he said. 'It was randomised, it was replicated. Everything about it was scientific. You can trust the science. If anyone wants to suggest we didn't catch enough badgers, it irritates me. That trial was the most rigorous that could have been devised. It was probably the largest field trial of its kind anywhere in the world. To rubbish the results is just not on. Paul Caruana was only involved with the field implementation of the trial. He is not a scientist and clearly does not understand or appreciate how and why the RBCT was designed the way it was.' The RBCT scientists calculated they could trap up to 70 per cent of the population in short, intensive trapping periods and that extending this beyond twelve days would catch more 'immigrant' badgers and disrupt the wider badger population even more. Professor Rosie Woodroffe, another badger expert who worked on the RBCT, described it as 'more or less military' in its implementation.

The flaws in the RBCT would crop up in any cull. The constraints on finding and killing every badger in a sett – protesters, the weather, animal welfare considerations – are universal, in Britain at least. The only guaranteed way to kill more badgers would be to use poison, and this would risk the accidental slaughter of various 'non-targets' such as foxes, otters and family dogs. Farmers still talk about humane forms of gassing, but to the general public no form of gassing is humane. Some setts are so labyrinthine that no gas would be completely effective. Gassing badgers is currently illegal in Britain, while wiping out entire local populations, as noted earlier, could contravene the Bern Convention.

I was interested in Tim Roper's view of the trial because he was not directly involved. 'It's the best piece of science that could have been done in the circumstances,' he said. He surprised me, however, when I asked if Bourne's 'no meaningful contribution' conclusion had become less valid with the emergence of post-cull data showing that culling did have a long-term impact on reducing bovine TB in cattle. 'I never thought it was valid,' said Roper, who believed Bourne's recommendation that policymakers should focus on the cattle-to-cattle transmission of the disease was 'strange' considering the RBCT's results, which appeared to show that badgers could rapidly pass the disease to cows.

The only other European country with a significant bovine TB problem is Ireland. Here, culling has been extensive (and the Irish government survived a challenge that it had violated the Bern Convention). The Irish seem less bashful about killing *Meles meles* and more certain it works. The Four Areas badger cull there saw an overall reduction of bovine TB in cattle of between 50 and 75 per cent.

Why did culling work better there? Many factors were unique to Ireland, a society which probably valued its farming more and its badgers less: more landowners agreed to take part, there was no public opposition, there was a commitment to eradicating badgers from a third of the countryside, and the creatures were caught with snares, a more reliable killing method than trapping. Some of the disparity, however, may be due to different behaviours not in humans but in badgers. Irish badgers could play a bigger role in spreading bovine TB than British ones and so their slaughter has a bigger impact. Ireland is now undertaking reactive culling. Badger lovers fear this will see significant population losses across the country. As Roper wrote, drily: 'Ireland does not seem much interested in cattle vaccination as a control option.'

The proposed 'farmer-led' cull in England was, in some respects, a cunning political manoeuvre. The government gave itself a huge problem when it assumed responsibility for killing badgers in the 1970s. Allowing farmers to cull them gave the farmers a welcome sense they were being listened to and shifted responsibility onto their shoulders. The state would restrict itself to a regulatory role. Most attractively of all, it saved the government money. Farmer-led culling was a bargain, costed at £300 per square kilometre for shooting free-ranging badgers versus £2,500 per square kilometre for trapping them in a cage first before shooting them. If badgers were living too near humans to be shot safely, they would be cage-trapped, but most of the cull would be cheap 'controlled' shooting.

For cynics who doubt whether science ever informs political choices, the modest changes in the data obtained from the RBCT provided a convenient argument for Conservative politicians who had always been determined to cull the badger. The coalition insisted its cull was 'science-led', saying the average net benefit of a 12–16 per cent reduction in new cases of cattle TB after five years of culling and a four-year post-cull period was equivalent to preventing forty-seven out of 292 cattle herds succumbing to TB in the culled area and its two-kilometre periphery. Arguably, the farmer-led cull might be called a common-sense cull, but this method of killing badgers could not be described as scientific. While ministers continued to insist they had science on their side, the department's mandarins appeared to concede the point. As a Defra document justifying the cull concluded: 'It is a matter of judgement, not science, whether the farming industry can deliver an effective, coordinated and sustained cull.'

Unfortunately the cull was unlikely to broaden our understanding of how bovine TB could be reduced in badgers or cattle. The RBCT showed the effectiveness of trapping and shooting badgers but did not provide any data on the efficacy of simply shooting them. Shooting could cause even more perturbation. And indeed, the farmer-led cull was not scientific: there were no control areas and the two zones were too small to permit statistical analysis. Also, although there would be some post-mortems to check whether badgers had been cleanly killed, there would be no comprehensive examination to record what percentage were carrying bovine TB. Defra considered that plenty of data on the (low) incidence of TB in badgers had been collected under the RBCT. Opponents of the cull, however, saw this refusal to test thoroughly as immoral; healthy badgers would be killed and critics believed the government was wary of generating data that would show what a small proportion of shot badgers were actually diseased.

Tim Roper, I learned, was to sit on an independent panel of experts who would assess the efficacy, safety and humaneness of the cull for the government. This link with Defra led me to expect him to be guarded in his opinions, but he was attractively frank. Looking at forty years of policy failure over bovine TB, Roper felt it had been a major tactical mistake to gas badgers in the 1970s. 'There was a lot of hostility to that – some reasonable, some unreasonable. Had farmers been licensed to shoot badgers at that stage there probably wouldn't have been a problem at all,' he said. Roper understood the political attractions of the current cull, and believed it would not be 'a free-for-all' on the badger population, but he was highly sceptical that it would reduce cattle TB. 'If you are really committed to culling (as the government seems to be), then the most effective way to do it would be systematically. I'd also be much happier if they said, cull for five years and then vaccinate the relic population,' he said. Will culling work? 'It depends what you mean by work. If enough badgers are killed, it will produce a temporary and slight decrease in the incidence of TB in cattle. If you think that's "working", culling works. But that's not what anyone would call a cure.'

*

Brian May's small gesture at the closing ceremony of the Olympics was the starting-pistol for wider hostilities. For a few weeks in the autumn of 2012, the furore over the badger cull caused a surreal shift in reality. People, organisations and companies with no connection to life inside a sett suddenly developed a badger-consciousness. In urban Hackney, children at one primary school organised a petition to 'stop killing the badgers'. Activists threatened boycotts of supermarkets that bought milk or cheese from pro-cull farmers – Morrison's said it would continue to buy from farmers on whose land badgers were killed; the Co-op promised it would not. Every celebrity was required to form an opinion on the cull: Dappy, the rapper from N-Dubz, was perhaps the most unlikely candidate to support the badgers.

The human and beastly inhabitants of Badgerland had been thrust, blinking, into the mainstream. Farmers perched on breakfast television sofas in suits, looking as uncomfortable as Ted Burgess, the farmer in *The Go-Between*, when compelled to wear a lounge suit and starched collar. 'The more clothes he put on, the less he looked himself,' it was said of Burgess in L. P. Hartley's novel. Against them, usually, was pitted Brian May, who became the badgers' PR man in TV studios across the land. May created a new coalition to oppose the cull which he called 'Team Badger'. Most of May's team were familiar exponents of rights for animals – the RSPCA, the League Against Cruel Sports, Humane Society International UK, Born Free and PETA – but they also included Conservatives against the Badger Cull and even Druids. 'Humans say, "nature is red in tooth and claw". And it often is. But we do not need to be brutal,' proclaimed the Druid Network. 'It behoves us, as Druids, to stand up when unnecessary brutality is threatened – to walk through the battle and bring peace. And so, the Druid Network supports Team Badger.'

In the weeks before the cull was due to start, May publicised a petition against the cull on the No. 10 Downing Street website. The government was obliged to 'consider' any such e-petition for debate in the Commons if it attracted 100,000 signatures. As the 'stop the badger cull' petition crept inexorably towards six figures, another star of the omnivoreshambles made his presence felt. Fortunately for satirists, the anodyne Environment Secretary Caroline Spelman had been replaced by a much more flavoursome character.

Owen Paterson grew up on a farm in Shropshire. Unlike his 'completely urban and completely clueless' Labour predecessors in Defra, the MP for North Shropshire declared he was a man who understood the countryside. 'And I eat meat,' he told *Farmers Guardian* on his appointment. Paterson's time in public school, at Cambridge and working for his family's tannery firm failed to erase his strong feelings for the land, and he kept horses, chickens and Black Welsh Mountain sheep at his Grade II-listed country home where he lived with his wife, a viscount's daughter. Best of all, as a boy, he had tended Bessie and Baz, two orphaned badgers. 'I was perhaps about ten years old when a local farmer rang us up to say he had found a young badger and would we take it in,' Paterson once revealed. 'So we did; it was a female called Bessie and she lived in the boiler room. She was extremely intelligent, had a very low opinion of cats but loved the dogs. She was pretty well trained, she went in the car. Then we had another badger and as soon as they got together they dug their way out and ran away. It is terribly hard when welfare people see me as anti-badger, having known a badger very well.'

Paterson may have been a badger hugger as a child but he made his name accusing the Labour government of ignoring the animal's role in bovine TB. In 2004, as the shadow environment minister, he tabled around 600 parliamentary questions – a record – on bovine TB, which made him appear both deeply committed to finding a solution and bearing a slightly weird grudge against Bessie and Baz's wild descendants. Paterson quickly gained a reputation within Defra for being 'more gung-ho' than his predecessor. Increasingly abrasive as the omnivoreshambles intensified, he spoke of needing to 'bear down on wildlife' carrying bovine TB. Finally, in the five-hour parliamentary debate called as Brian May's petition soared beyond 150,000 signatories, Paterson could only bear to stay for twenty minutes, and resorted to bellowing, rather like a wounded animal, at the MPs baiting him. Democracy backed the badger: MPs voted decisively in favour of the motion to abandon the cull – 147 to 28 against – although this would not be binding on the government.

For all of Paterson's star quality, and May's celebrity, both were eclipsed by an even more charismatic character. 'The badger,' said one commenter on the *Guardian*'s website, where an ill-tempered debate also raged for days, 'is the tutelary deity of England. Destroy them at your peril.' It was an excellent choice of words. During the culling crisis, the badger came to be seen as the guardian of the countryside.

Stopping the cull was a simple way for millions of people to feel they were doing something right by their beleaguered wild places.

The badger was the winner because it was both native and exotic. We imbue it with integrity, and see it as quintessentially British because of its long presence on our lands, and yet we do not take it for granted because most of us rarely see it. No matter how hard farmers try to persuade us, *Meles meles* will never be viewed as a pest like the rabbit, magpie or rat. For all its symbolic resonance, the badger's visual qualities should not be underestimated. No matter how bad its crimes, no jury would ever find such an appealing criminal guilty. The images that accompanied every cull story in newspapers and on television showed a healthy badger with a misty gaze as querulous and innocent as a short-sighted old lady. There is also something intrinsically amusing about badgers. The comedian Matt Lucas once talked of stand-ups who could inspire laughter simply by the way they looked and talked; he called it having funny bones. Badgers have funny bones too, and are reliably deployed by comedians from Harry Hill to Marcus Brigstocke.

During the cull crisis, badgers were regulars in Steve Bell's *Guardian* cartoon strip, being shot by the Queen (who called them 'bedgers') and becoming Prince Charles's new beefeater-style hat. One of the most powerful sketches of all appeared in Defra's 'best practice guidance' for the 'controlled shooting of badgers in a field'. Two line drawings of a badger showed where it needed to be shot so as to be cleanly killed, and how easily a 'robust bony limb' could intrude over the target area, making a clean kill deceptively difficult. This hapless demonstration badger should have been rendered with an evil grimace but, unfortunately for the government, it only looked cute.

These images begged the question: would a shotgun administer a quick and painless death for a badger? Chris Cheeseman shot badgers as a teenager, when it was still legal to do so. He used to visit a farm in Dorset where badgers were taking free-range chickens. One night, using a shotgun with triple-A shot, he dispatched three in the farmer's yard. Shooting in the field, however, was difficult. He would kill one and then hear a scuffling as every other badger retreated to the sett. 'It was highly disruptive,' he remembered. 'I thought, this is a waste of time. It was like taking a handful of beans out of a jar – there were still plenty left.' Shooting, he knew from personal experience as well as scientific evidence, would be a recipe for perturbation.

Farmers, however, were quite sanguine about dispatching badgers. Evan Thomas, the old Welsh dairy farmer I spoke to about bovine TB, said he too had shot badgers when it was legal. 'You shoot a deer from 200 to 300 yards. You'll be shooting a badger from fifty to sixty yards,' he said. 'At that distance, any competent rifle user should be able to hit a 50p coin.' Did he ever maim a badger? 'If you injured one you fired the second shot immediately afterwards to make sure you did your best to kill it,' he replied. 'You don't set out to injure an animal.'

While Brian May orchestrated the popular opposition to the cull, a growing number of scientists added gravitas to the antis' arguments. Lord John Krebs called the cull 'mindless'. Professor John Bourne, who ran the RBCT, warned it might make bovine TB 'a damn sight worse'.

Perhaps most damagingly for the government, its own chief scientist John Beddington left Defra dangling with this beautifully diplomatic gem: 'I am content that the evidence base, including uncertainties and evidence gaps, has been communicated effectively to ministers.' It was impossible to find a scientist who publicly supported the government's farmer-led cull.

The most significant intervention came when a group of eminent scientists, a *Who's Who* of international biology and ecology, wrote a joint letter to the *Observer*. Signatories included Krebs, Bourne, Cheeseman, Professor Stephen Harris, Professor Rosie Woodroffe, and other eminent experts in wildlife disease with professorships everywhere from Princeton to Cambridge. These scientists reiterated the accepted scientific view 'that licensed culling risks increasing cattle TB rather than reducing it'. Badger culling, they warned, could become 'a costly distraction from nationwide TB control'. As their carefully worded conclusion put it: 'We recognise the importance of eradicating bovine TB and agree that this will require tackling the disease in badgers. Unfortunately, culling badgers as planned is very unlikely to contribute to TB eradication. We therefore urge the government to reconsider its strategy.' This was not as emphatically against killing badgers as it first appeared. The scientists were not ruling out every kind of cull, only 'culling badgers as planned'.

The group closest to the popular mood were those who were implacably opposed to any violence against badgers: the animal rights activists. At first, there did not appear to be many people prepared to take direct action to stop the cull, but they made a big noise. The culling companies were intended to preserve nervous farmers' and shooters' anonymity, but a maverick website, Badger-Killers, published names and addresses of the company directors. The leading light in GlosCon, the culling company for the Gloucestershire hotspot, was revealed to be Jan Rowe, a dairy farmer I had met. Jan seemed a decent man and now he and his wife were bombarded with crank calls and threatening texts. Malicious reviews of their bed-and-breakfast business were also posted on TripAdvisor by anti-cull protesters. When photographs of Owen Paterson's home and his office phone number were posted on Badger-Killers, government lawyers threatened an injunction and the addresses were removed. But it was too late to stop protesters plaguing Paterson's Westminster office with calls in which they played an exceedingly irritating novelty song that went 'badger-badger-badger' down the line. Inevitably, another website appeared, hosted in a foreign country, and continued to publish the addresses of those involved in the cull.

This was armchair direct action – annoying and sometimes threatening emails and phone calls by activists otherwise busy with their day jobs. As animal rights veterans acknowledged, they could no longer rely on a 'doley army' as they had back in the 1980s. Nevertheless, they arranged to bus people to the cull zones at weekends, and planned nightly patrols to disrupt the shooting. Gloucestershire police warned that they could, ironically, detain protesters under the Protection of Badgers Act 1992, if the activists disturbed the animals around their setts. But, despite these threats, it looked remarkably easy to legally disrupt the cull. As long as they could find public footpaths near the targeted setts, activists could scare badgers away from the gunmen with torches, whistles, fireworks and those vuvuzelas. A more imaginative tactic included pouring circles of human urine around any feeding stations. Weirdly, only male urine would work, not female. 'The only major problem we have with this is that the female to male ratio is about 2:1,' explained one activist. Ultimately, the battle between farmers and the animal rights movement would be about propaganda: the holy grail for activists was gruesome footage of badgers being shot dead; both sides knew this, and TV crews and journalists were told they would not be given access to the shooting.

Rumours of more violent direct action spooked the farming industry. Farmers reported that activists were threatening to plant Japanese knotweed, a particularly virulent alien plant, in their fields. Worse, farmers in the cull zones feared their barns or tractors would be vandalised or burned by extremists. Some months before the omnivoreshambles, Stewart Thomas, a dairy farmer in the Somerset cull zone, spoke in favour of the cull in his local newspaper. Ten days later, his barn burned down.

I went to see Thomas at his elegant farmhouse, surrounded by a walled garden filled with ordered rows of vegetables grown by his wife, Marie. He had short grey hair and a limp from a bad back and, without trying, soon displayed the keen eye for wild animals that comes naturally to people who work on the land. He saw a 'little sow' badger quite regularly on the lanes. 'She has a lovely shiny coat,' he mused. One day, he was driving behind her 'and she was looking back as if to say, "Don't you run me over." I think they are quite intelligent.'

Thomas tended 500 dairy cattle and, despite taking measures to stop badgers coming into his yards and farm buildings, he was currently shut down because bovine TB had been detected in his herd. One farm not far away had a herd of a hundred sucklers in which seventy had reacted to the TB test; the vet's theory was that a dead badger with TB had been picked up in the harvester and rolled into silage fed to the cows. For Thomas, now arson had been added to the trial of being shut down. The lad who milked in the mornings saw the glow. The first blaze was lit after 3 a.m.; another day, a fire was started shortly after 4 a.m.; a third act of sabotage saw his milk contaminated with chemicals. The police would not formally link this vandalism to the farmer's public endorsement of the cull, but Thomas could not 'see any other reason. You can upset somebody and they will let your tyres down and you do get arsonists but they wouldn't come back to do it a second time, and they wouldn't contaminate the milk.' The arsonists used a delayed device to start the fire, which suggested they knew what they were doing. Because he was under-insured, Thomas calculated he had lost £15,000 in the attacks.

As farmers' fears grew, more spoke out against the cull, sometimes arguing on pragmatic grounds that it would be a PR disaster for British agriculture.

After the first licence was issued to the GlosCon culling company in the late summer of 2012, it seemed a matter of days before the cull would start. Each week, however, a new obstacle appeared. Paperwork was shuffled. The firearms licences of the sixty shooters in each area tasked with bagging the badgers (paid at £50 a carcass, a decent-looking inflation-adjusted update of the medieval bounty of twelve old pence) had to be amended. Then it was revealed that despite months of preparations, no up-to-date sett surveys had been undertaken. The government needed to know how many badgers lived in the cull zone if it was to judge whether the cullers had shot the requisite 70 per cent. Fifty-five officials were dispatched to survey setts in the area. They also took DNA samples from badger hair caught in coiled pieces of barbed wire placed above well-worn badgerways, data that would be used to monitor the effectiveness of the cull. (Photos of these DNA traps were published on Badger-Killers, which urged activists to sabotage this important means of information-gathering.)

Defra revealed that the costs of the cull now outstripped the benefits. Farmers were projected to benefit by £3.68 million, while the basic cost of the cull was £4.56 million with an additional £4 million police bill over four years. These cost–benefit calculations appear to be rational but will always be shaped by the values and priorities of policymakers. In reality, no one could sensibly estimate the price of policing the protests because no one knew what those protests would be. And in Defra's calculations, revealingly, neither badgers nor the affection in which they are held were given any economic value.

As these flustered final preparations were made, the pressure against the cull continued to build. September turned into October, and still the cull had not begun. The badger cullers were running out of time. They needed a six-week window before the close season: cage trapping was forbidden beyond 1 December, and shooting beyond 1 February because of the risk of pregnant or suckling sows being killed. Whenever anyone asked, officials in Defra, or Natural England, or the NFU, would say there were just a few tiny details left to sort out. Those tiny details, however, sank the cull.

In the Commons, the government was forced to admit its last-minute census had found 3,600 badgers in the Gloucestershire cull area and 4,300 in the Somerset zone. This was double the previous estimated number for the Gloucestershire zone and 60 per cent above an earlier estimate for the Somerset one. It would make the cull more expensive (with the bounty) and more difficult to meet the 70 per cent killing target. At the last minute Natural England, which awarded the licences for the cull, decided the target must actually be to kill 80 per cent of badgers in each zone.

With time, money, the prospects of success and the public mood increasingly against them, farmers were by now, behind the scenes, begging the government to call off the cull. In mid-October, days before the debate in Parliament triggered by Brian May's petition, Owen Paterson did just that, and postponed his precious cull until the following summer. The omnivoreshambles was over, for now.

These febrile few weeks perfectly reflected the confusion, indecision and conflicting evidence from the previous forty years. It appeared that we had learned nothing, and decided upon less. Cattle continued to be slaughtered because they tested positive for bovine TB and badgers continued to be blamed. The usefulness of killing badgers to reduce disease in cattle had not been scientifically established. The only change had been a massive surge in the public's apparent affection for the badger and the only sure thing was that *Meles meles* would have another peaceful winter. Of course, they would continue to be hit on the roads, illegally shot, dug, baited, poisoned, snared and have slurry poured into their setts, but there would be no state-sanctioned slaughter of our badgers, not quite yet.

16

Not My Badgers

The Norfolk countryside was calling me. I had finally moved to Norwich, from where the countryside of my childhood was only a thirty-minute drive. For the first eighteen years of my life, I had roamed a valley by a small stream we called Booton Beck. There was a common, a poplar plantation and a few rough water meadows. Although boggy, most of the land was ploughed up during the apotheosis of industrial agriculture in the 1970s. The two animals my mates and I most hoped to see were coypu and the Beast of Booton. Coypu, exotic, beaver-like rodents, had been an accidental introduction and were on the brink of being successfully exterminated once again. The Beast of Booton was a lynx – or more than just one – which had probably escaped from the local wildlife park. Most people assumed the Beast was mythical but some years later, in the 1990s, a local gamekeeper suspected of shooting birds of prey was visited by police. When an officer asked what he kept in the large freezer in his outhouse, he replied: 'Oh, only some pigeons and a lynx.' He had shot one as it chased his dog across the fields a mile from my childhood home.

This was a landscape bereft of native fauna. There were no exciting indigenous animals: not much chance of seeing a deer; no buzzards or marsh harriers or even barn owls; few hares, no otters and certainly no badgers. East Anglia has probably never sustained a badger population as large as the West Country's because they cannot build setts on flat, poorly drained land; but the rise of the big estates and a century governed by the gamekeeper followed by the rule of prairie fields exterminated most of the badgers that did live here, as well as any beast much bigger than a rat.

When I returned to the countryside around Booton and the small town of Reepham, my childhood stomping ground, I was in for a tremendous shock. On a sunny spring day, I bumped along the old railway line with Lisa and our buggy-load of newborn twins and wondered what kind of weird bewitchment had taken place in the decade I had been away. Acres of arable fields along the valley bottom had been returned to pasture. My former neighbour had transformed his farm into a conservation concern, his natural love of the land encouraged by the European Union's environmental stewardship schemes. A buzzard nested in the spinney. I ran into another old neighbour who said that otters were back on the beck and I soon found the slides down which they joyfully plummeted into the water. Booton Common was thick with traces of fallow and sika deer as well as the ubiquitous muntjac, and there was also a weird scratch mark high on a trunk that could have been scored by a big cat. The granddaughter of the Beast?

Given that the once derelict railway yard three fields from my former home now housed working steam trains, it appeared that this present was a weirdly warped version of a more distant past. We walked and wondered and I bemoaned the fact that despite this renaissance in pasture, wildlife and steam trains, we had still moved to the worst place in Britain to find badgers. At that very moment, Lisa, the twins and I found our footpath blocked by a new steel gate. A sign announced this section of the railway embankment had been closed because of 'badger activity'. It was deemed unsafe, undermined from below by badgers.

I clambered over the gate and made my way along the forbidden path. The council were fussing over nothing – the path had no cavities whatsoever. Then I looked over the side of the embankment. There, on the west-facing slope, was one of the biggest setts I had yet seen, more than twenty crater-like holes each spewing out a mini-mountain of sandy soil. It was epic, thrilling, and I also felt sad. My old house was just over the horizon; my school, one field away; our cross-country route passed right by here. I could've spent my schooldays badger watching and I never got the chance.

Badgers were back in Norfolk. As if to confirm the fact, I started seeing them dead on the roads, where I had never seen them as a boy. For weeks after visiting the fields around my childhood home, I felt haunted by the gulf between my old memories and the place of the present day. Going back is always unsettling. We usually experience return as a loss, noticing where wildlife has been extracted and new buildings or roads inserted into landscapes that were pristine in our memory. The countryside of my childhood, however, had been enriched, not denuded, filled with new hedges, woods, footpaths, and mammals. It was wonderful, and it was unnerving.

I had chanced upon the vast badger sett on the old railway line and had to go back.

A miserable spring at least held the blooms of April in flower for a month and fields of oilseed rape still shone brilliant yellow while cow parsley filled the field banks with latticed cream. The youthful ash trees on the railway embankment undermined by the enormous badger sett were just coming into leaf. I ignored the notices warning of hazardous badger excavations and vaulted the steel barriers blocking the closed footpath. The sun was still high but the badger watcher's mantra – be there before sunset – was now ingrained in me. The ashes were too slender to climb and there was no other cover on the embankment so I plunged into the valley bottom it cut across. A boggy triangle of pasture fenced off with wonky barbed wire and left to its own devices now bore the most perfect crop of nettles. A neater field could not have been sown. The nettles were waist high, and in peak condition; fecund and poised to flower, not dry and stringy as they would become at the height of summer. I breathed in that nettle tang: a bittersweet reminder of times past.

This was a field of a million stings, and they even jabbed through my jeans as I waded among them. A muntjac fled ahead of me, rippling through the nettles. At the far corner, eighty yards from the badger sett, I sat down in them, creating a deer-like dell. Seated, my eyeline was level with their tops, giving me a nettle flower's view of the world, a casual intimacy with the natural world I last had in the days when I made hideouts on the overgrown meadow by our home. Being here was the most soothing form of meditation I could imagine. I savoured it, a break from a life of constant phone calls, travel, London, and now the intensity of the twins.

I was unfamiliar with waiting for badgers in tepid conditions, and the warm evening brought out another new experience: insects. I had not realised how insectless the early spring had been. Now they were everywhere. Up my leg crawled a pale emerald beetle. Hundreds of midges hung like Chinese lanterns in the sky. Others swung over nettle heads, motes of dust with purpose, moving as regularly as pendulums. The mosquitoes were obviously not expecting a warm-blooded lump to settle in the nettles – it took fifteen minutes for the first biting insect to find me.

I had a fine view of the badger city laid out on the western edge of the embankment. Mounds of sandy earth scraped from the tunnels erupted among the greenery. But I still doubted I would ever see a badger here. They had always been extinct in the landscape of my childhood. And so far, the embankment was populated only by some grand rabbits the size of coypu.

There was a pheasant, a wood pigeon and a bird that sounded like the whistle of a man. Then, a mile beyond, a cuckoo, a sound made melancholy by its rarity compared with the plenty of years ago. From a ditch of uncertain depth behind me I could hear the whirring wings of a small bird. Something else, mammalian, trod the damp leaves in the ditch bottom. A big sound from a shrew-sized thing. The predators knew the prey were abroad, for a tawny owl then swept from the wooded embankment like a low-flying alien craft. It flew just above the nettles, its eyes fixed firmly forward and its great satellite dish of a face tapering into nothingness behind.

Dusk was a mixing of light, temperature, past and present. Cold air sank into the valley, blending with warmer air that swirled the smells around. Brightness in the western sky met shadow and softened. Creatures of the day encountered creatures of the night. Movement saw stillness. Sensations mixed with memories, present with past, and as the singsong voices of children rang out from a campsite across the valley I remembered those magical summer evenings when my sister and I were allowed to play outside until way past our bedtime, when all the cuckoos in the valley called and scores of swifts screamed in the sky.

I was not really badger watching at all. I was dusk watching.

I stayed there, inebriated by the twilight, struggling to reconcile this wonderful and disorientating experience of seeking badgers in the terrain I knew best with the grim prospects for the badger in the wider countryside. We had failed to find a solution to bovine TB and badgers for more than forty years and we appeared to be no closer than we had been in the early 1970s. This failure was because people could not agree on the badger's place in our landscape. Was it vicious or benign? Diseased or cleanly? Cherished symbol or vermin? Rare or common? Should it be completely protected like, say, the similarly resurgent red kite, or managed, hunted and culled like red deer?

I had been sitting on the fence for months because I wanted to give a fair hearing to all sides but also because I, too, had no miraculous solution to the problem of bovine TB. It wasn't even that: I had been looking for an argument I believed in. No one entirely convinced me; not the badger lovers, nor the badger haters. Even the scientists were limited by their frames of reference. I was not confident that we would ever introduce a badger cull brutal enough to have an impact on TB. Nor did I think our policymakers would ever decisively renounce the crazy proposition that to farm successfully we need to kill large numbers of a native animal. Badgers would always be fair game as long as civil servants arbitrarily decided their economic value to us was zero. A cull coordinated by the government, as the supposedly farmer-led cull was, would always be peculiarly offensive, smacking of an industrial-scale cleansing, but I felt sorry for hard-pressed individual farmers, and wondered why sensible countrymen could not be allowed to quietly dispatch problem badgers with a shotgun, just as they did with foxes. That would be a true farmer-led cull, and farmers would then have to deal with any unwelcome consequences of perturbation.

Our attachment to badgers was not rational and there was a reasonable argument for saying we should not treat them any differently from rabbits or rats, which lack the badger's special legal protections so fought for by my grandma. Unlike animal rights activists, who tended to believe all species should be treated the same, I instinctively sided with the conservationists and ecologists who argued we must treat different species differently. The animal rights movement called this 'speciesism' and likened it to human discrimination against different nationalities, but it is not fascism to cull grey squirrels we brought to Britain when they threaten the existence of our native red squirrels, or to cull introduced species of deer when they destroy our woodland flora and the butterflies and other insects dependent upon it. 'Conservation cull' is not an Orwellian phrase; rightly or wrongly, we have appointed ourselves custodians of the natural world and it is incumbent upon us to seek to preserve its balance as best we can, especially when we have already disrupted and destroyed so much.

But a cull of badgers *is* different from killing rats or grey squirrels: they are a key species in our ecosystem, they are not an introduced species, they are not yet of plague proportions, and their predation of ground-nesting birds, bumblebees and hedgehogs, although significant in some areas, is probably exaggerated. The only things they seriously threaten are the rather weak cows we farm intensively and make susceptible to disease. Bovine TB in cattle might just be reduced to an unproblematic level by vaccines for cattle and badgers, although there are plenty of people who remain sceptical that vaccines will fix the problem. Farming, the unnatural concentration of living organisms, will always generate disease; bovine TB exists in our environment and will mutate into new forms.

Late one evening, I was flicking through debates about culling when I finally found an argument that was utterly convincing. To my surprise, it lay in the pages of a veterinary journal. Most of the veterinary industry supported the badger cull. But here, in the *Vet Times*, a group of six vets offered a powerful case against the whole way in which badgers had been drawn into the argument over bovine TB.

A vaccine for cattle could only be a sticking plaster, they argued, when the fundamental problem was the evolution of the dairy industry. Artificial insemination (AI), widely used since the 1950s, had selectively bred 'mutant cows' that produce large quantities of milk but have little resistance to diseases such as bovine TB and BSE. 'Infectious agents and their hosts tend to adapt or co-evolve together such that a balance is formed between infection, immunity and survival, and this is demonstrably true for TB. In badgers, this balance with TB has happened over millennia. Dairy cows stopped co-evolving with TB more than 50 years ago, due to AI. The only thing dairy cows have co-evolved with is human will, industrial economic policy – and money,' the six argued. The agricultural industry, they suggested, should breed dairy cows with new stock to return some 'hybrid vigour' to these unfortunate animals. In a brilliant passage, the vets concluded:

TB is often a disease of poverty, in humans as well as animals, and many of our dairy cattle live in poverty equivalent to that of a workhouse during the industrial revolution. Most importantly, there is poverty in the lack of any normal relationships around breeding and calf rearing. The only long-term solution is a paradigm shift in favour of cattle welfare, small farmers and wildlife – not mega-dairies and money. We need to start looking, right now, at the economic and genetic background to the dairy industry, and fix it, before it's too late. We support the long-term restructuring and de-intensification of the dairy industry to better support the health and welfare of cattle, as well as that of small farmers and consumers. This would go some way to help to ensure a more natural, less pressured life for the dairy cow.

For an idea of what this argument might look like in real life, I talked to Steve Jones, an influential cattle manager who had become a farm consultant. 'We could bring cattle vaccination in tomorrow and the government would still be looking at a badger cull programme because vaccination wouldn't help, with the British farming industry being in the state it's in,' he argued.

Jones was not putting the case for a badger cull, however, but for better animal husbandry. Most bovine TB-affected farmers were infuriated when conservationists suggested that the disease could be solved by tighter biosecurity – fences to keep badgers away from farms and suchlike – because they were already doing all they could, but Jones made a more far-reaching set of recommendations. As the vets argued, bovine TB is a disease of poverty and a range of afflictions – BSE, foot-and-mouth, dermatitis and TB – have arrived in an era when small British farms are impoverished and dairy cattle in particular have been weakened by selective breeding. Looking at the old-fashioned dairy shorthorn cow next to a modern Holstein is 'like comparing Land Rovers with Ferraris,' said Jones. The high-performance Holstein may produce more milk but it is far less robust. It is also a much bigger animal, and many old-fashioned farms don't have large enough beds, sheds or loafing areas for their livestock. Struggling farmers cannot afford to address overcrowding, muddy yards, poorly ventilated sheds, the regular cleaning of water troughs (a potent potential transmitter of bovine TB, according to Jones) and the trimming of feet (22 per cent of cattle are 'lame', according to Defra). All these elements increase stress and an animal's susceptibility to bovine TB.

Jones also challenged my assumption that the search for higher productivity through industrial farming was the root of many evils. During his career, he had advised mega-dairies in Saudi Arabia, where milking units contained 5,000 cows. Here, he saw good disease resistance among high-yielding cattle, kept in appropriate conditions with specialist staff. Some units would even have a full-time vet. 'These animals have all their needs met. They are better looked after than on a small, cash-strapped family farm,' he argued. Jones was not advocating bigger or smaller farms for Britain, just better farms, which might take shape if dairy farmers were given a fair price for their milk.

Such radical changes in animal husbandry and selective breeding were idealistic and probably unrealistic. They would require a completely new kind of agricultural ministry and unprecedented intervention in the free market. They would make food more expensive and would probably require us to leave the EU. But they would return real power to the individual farmer in the countryside, rather than the phoney semblance of autonomy granted to them in the form of a 'farmer-led' cull of badgers. Such an agricultural revolution would liberate farmers, and badgers.

These rational-sounding answers to the question of bovine TB were

soothing but they concerned farms, not badgers, and did not acknowl-
edge the strange and contrary feelings our species has for *Meles meles*.
We just cannot leave the badger alone. We are compelled to find it,
watch it, feed it, photograph it, poke it, catch it, torture it, defend it,
kill it. Perhaps it is too big a mammal and plays too significant a role
in our landscape for us to ever leave it in peace. It is virtually a com-
petitor, or the nearest thing left to one on an island denuded of big
mammals. When its interests clash with ours, we seek to 'manage' or
exterminate it, just as we can't help ourselves in our curiosity, greed
and desire to seek dominion over every other part, no matter how tiny,
of our miraculous world of animals and plants.

Yellow-tinted clouds shifted slowly overhead. Cooler air rolled down
my forehead, bringing with it the cabbagy pong of the oilseed rape in
the field beyond. This was not good news because it meant I had mis-
calculated the wind direction. If I could pick up this brassica breeze
when I had a nose full of nettle, then the badgers would certainly
detect me, even though I was eighty yards away. The Norfolk badger
recorder (there's one in every county) told me how badgers could sniff
out a stranger from 400 yards. These animals were the ultimate
sommeliers, trained through the generations to sniff, and know, everything, including my careless scent.

But shortly after 9 p.m., a flash of white showed between two
humps of nettles on the embankment. A nose, gone before I realised
it had arrived. A few seconds later, a grey beast trotted springily up one
of its paths to join the white nose. Here were the Norfolk badgers,

who had thrived in the years I had been away and made this their home. There were no more coypu, and the Beast of Booton was as elusive as ever, but these alien species had been replaced by something far better, our biggest native carnivore, the tank of the woods, who had lived in East Anglia before man arrived and would probably still be here long after we departed.

As this pair raced around the embankment together they reminded me of the way my twins were now crawling at high speed, rolling and bumping around without ever upsetting each other, and I felt a pleasing sense of belonging to the mammalian world. Then, just as suddenly, the badgers disappeared. They had almost certainly sniffed out a suspicious trace of me.

Two bats. Dot, dot, against the sky. I could see only silhouettes now. A heavy calm settled on the marshy field, except for a noise in the distance, the soundtrack of my youth, the hairdryer lilt of a 50cc motorbike straining up the hill.

I saw no more of the badgers that night. If I had a claim to any group of badgers, then surely it was this cete. I was home and I had found my Badgerland but I did not feel for a minute that these were my badgers. Badgerland is a kingdom that we have created, full of terror and joy, from the bad men who respect the beast they torture to the good people who inflict an ignominious subjugation on a wild animal, as John Clare showed in his poem, quoted earlier, when he wrote of a miserable badger as tame as a 'hog' that would follow its owner 'like a dog'. Badgers were not to be neutered like this. And I would be foolish to form any kind of imagined relationship with these badgers, even though I encountered them close to my childhood home and knew that this was where I would seek them out again, and hopefully bring my twins too, whenever we felt the urge to dive into the calm, inexorable magic of dusk.

I was tired as I climbed back into the car but my senses felt more alive than ever. Badger watching, dusk watching, was where beings of the day met beings of the dark and both types of creature were transformed. Shadows lengthened, sounds sharpened and memories were awakened. It could be a golden time, a gloomy time or a drowsy time and yet it was as vital as listening to music through headphones with your eyes closed in the hot sun; it was a warm bath, a wet run, a cold swim; all those greedily taken sensory pleasures. I had bathed in nettles, insects, birdsong and tepid air; the dusk had seeped into me. For a fleeting moment, I had become it, and that was enough for anyone.

ACKNOWLEDGEMENTS

I want to thank everyone who appears in this book and who generously gave up their time to talk about badgers. In particular, thank you to Judy Salisbury for her friendship and Nick and Sue Lee for their hospitality and help. I would also like to thank Ronald Blythe, Maureen and Charlie Davies, Michael Dougdale, Oliver and Jill Edwards, John Field, Robert Fuller, Don Hunford, Steve Jones, Julia Kaminski, Pauline Kidner, Jess Lee, Michael and Phyllis Lee, Jane Lewis, Gordon McGlone, Jonathan McGowan, Dave and Patsy Mallet, Brian May, Chris Packham, Jan Rowe, Evan Thomas, Stewart Thomas, Jean Thorpe and Jez Toogood.

A number of scientists and experts were generous enough not only to talk to me but then to read drafts and point out my mistakes and misunderstandings. Any that remain are my own. I am extremely grateful to Chris Cheeseman, Tim Roper, Chris Newman, Christina Buesching, Sultana Bashir and Amanda Barrett.

Thank you to Karolina Sutton, lovely and formidable as all agents should be, and her assistant, Catherine Saunders. Most writers think that editors are like cooks with broth but I was lucky enough to have two for *Badgerlands*. Sara Holloway and subsequently Laura Barber were brilliant, both with forensic detail and the bigger picture. Any inadequacies that remain are my own. Thank you to Jake Blanchard for his creativity and Sue Phillpott for her great care and attention. I also want to thank members of Granta for their dedication and professionalism: Iain Chapple, Philip Gwyn Jones, Christine Lo, Brigid Macleod, Aidan O'Neill, Kelly Pike, Pru Rowlandson, Michael Salu, Sarah Wasley, and the rest of the team.

Many other friends, relatives and acquaintances helped with ideas and insights, including people whose tips I remember ('Read Tom Jaine', for instance) but whose names I have forgotten. I'm sorry. Thank you to Amelie Barkham, Carla Barkham, Henrietta Barkham, Angela Cassidy, Sean Clarke, John Crace, John Crouch, Christl Donnelly, Ben Gaskell, Rebecca Gethin, Louise Gray, Robin and Rachel Hamilton, Henrietta and Hugo Kidston, Richard Mabey, Liz Norris, Stuart and Jenny Patterson, Paul Ratcliffe, Jennifer Rowland and John Sutherland.

A special thank you to my mum, Suzanne Barkham, for her memories and help, and to my dad, John Barkham, for his enthusiasm and generosity. Most of all, I would like to thank Lisa Walpole and our daughters, Esme and Camilla Barkham, for their love, support and inspiration.

SELECT BIBLIOGRAPHY

H. Mortimer Batten, *The Badger Afield and Underground*, H. F. and G. Witherby, 1923

BB [D. J. Watkins-Pitchford], *The Badgers of Bearshanks*, Ernest Benn, 1961

J. Fairfax Blakeborough and Sir A. E. Pease, *The Life and Habits of the Badger*, Foxhound, 1914

Margaret Blount, *Animal Land*, Hutchinson, 1974

Ronald Blythe, *Akenfield: Portrait of an English Village*, Allen Lane, 1969

Ronald Blythe, *Word from Wormingford*, Viking Penguin, 1997

Ronald Blythe, *At the Yeoman's House*, Enitharmon Press, 2011

J. C. Bristow-Noble, 'Hunting Brock the Badger', *Sussex County Magazine*, vol. 3, 1929

Fred Brown, *Badgers in Our Village*, Grafton, 1990

Norah Burke, *King Todd: The True Story of a Wild Badger; and of the*

Deer, Foxes, and Other Animals in the Forest Where He Lived, Putnam, 1963

Gordon Burness, *The White Badger*, George G. Harrap, 1970

Humphrey Carpenter, *Secret Gardens: A Study of the Golden Age of Children's Literature*, Allen & Unwin, 1985

Stephen P. Carter et al., 'BCG Vaccination Reduces Risk of Tuberculosis Infection in Vaccinated Badgers and Unvaccinated Badger Cubs', *PLOS ONE*, 2012 http://www.plosone.org/article/info%3Adoi%2F10.1371%2Fjournal.pone.0049833

Paul Caruana, Environment, Food and Rural Affairs Committee 2006, *Bovine TB: Badger Culling*, vols. I and II, Stationery Office, 2006

Angela Cassidy, 'Vermin, Victims and Disease: UK Framings of Badgers in and Beyond the Bovine TB Controversy', *Sociologia Ruralis*, 52(2), 2012

Keith Childs, *The Badger Diaries*, Bookworm Publications, 2010

Horatio Clare, *Running for the Hills*, John Murray, 2006

Michael Clark, *Badgers*, Whittet Books, 1988

Tess Cosslett, *Talking Animals in British Children's Fiction, 1786–1914*, Ashgate, 2006

Nicholas Cox, *The Gentleman's Recreation* [1677], EP Publishing, 1973

Penny Cresswell, Warren Cresswell and Michael Woods, *The Country Life Guide to Artificial Badger Setts*, Country Life, 1993

Jim Crumley, *Badgers on the Highland Edge*, Jonathan Cape, 1994

Roger Deakin, *Wildwood: A Journey through Trees*, Hamish Hamilton, 2007

DG(SANCO), *2011-6057: Final Report of an Audit Carried Out in the United Kingdom from 05 to 16 September in Order to Evaluate the Operation of the Bovine Tuberculosis Eradication Programme*, European Commission, Health and Consumers Directorate-General, Food and Veterinary Office, 2011

Christl Donnelly, 'The Duration of the Effects of Repeated Widespread Badger Culling on Cattle TB following the Cessation of Culling', *PLOS ONE*, 10 February 2010

Phil Drabble, *Badgers at My Window*, Pelham Books, 1969

Phil Drabble, *No Badgers in My Wood*, Michael Joseph, 1979

H. L. Dugdale, A. Griffiths and D. W. Macdonald, 'Polygynandrous and Repeated Mounting Behaviour in European Badgers, *Meles meles*', *Animal Behaviour*, 82(6), 1287–97, 2011

Gareth Enticott, 'The Spaces of Biosecurity: Prescribing and Negotiating Solutions to Bovine Tuberculosis', *Environment and Planning A*, vol. 40, no. 7, 2008

Gareth Enticott et al., 'The Changing Role of Veterinary Expertise in the Food Chain, *Philosophical Transactions of the Royal Society, B: Biological Sciences*, 366 (1573), 2011

Chris Ferris, *The Darkness Is Light Enough*, Michael Joseph, 1986

Chris Ferris, *Beneath the Dark Hill*, Swan Hill Press, 1995

Dr J. Gallagher et al., Memorandum to EFRA: Environment, Food and Rural Affairs Committee, November 2007, http://www.bovinetb.info/docs/Gallagher.pdf

Wyn Grant, 'Intractable Policy Failure: The Case of Bovine TB and Badgers', *British Journal of Politics and International Relations*, 11 (4), 2009

Peter Green, *Beyond the Wild Wood: The World of Kenneth Grahame*, Webb & Bower, 1982

Harold D. Guither, *Animal Rights: History and Scope of a Radical Social Movement*, Southern Illinois University Press, 1998

David Harcombe, *Badger Digging with Terriers*, Fieldfare, 1985

David Harcombe, *The World of the Working Terrier*, Fieldfare, 1989

Peter Hardy, *A Lifetime of Badgers*, David & Charles, 1975

L. P. Hartley, *The Go-Between*, Hamish Hamilton, 1953

Robert Howard, *Badgers without Bias*, Abson Books, 1981

Janni Howker, *Badger on the Barge and Other Stories*, Fontana Lions, 1984

Peter Hunt, *The Wind in the Willows: A Fragmented Arcadia*, Twayne, 1994

Tom Jaine, *Cooking in the Country*, Chatto & Windus, 1986

Richard Jefferies, *The Gamekeeper at Home, The Amateur Poacher*, Oxford University Press, 1978

Eleanor Kerr, *Hunting Parson: The Life and Times of the Reverend John Russell*, Jenkins, 1963

Pauline Kidner, *Life with Bluebell and Other Tales from an Animal Orphanage*, Robinson, 1993

H. H. King, 'Working Terriers, Badgers & Badger Digging', *The Field*, 1931

F. Howard Lancum, *Badgers' Year*, Crosby, Lockwood & Son, 1954

Sir Roger L'Estrange, *A Hundred Fables of Aesop*, Introduction by Kenneth Grahame, John Lane, 1899

Ralph Lewis, *Mister Badger: A Study of Badgers*, Research Publishing, 1976

Jocelyn Lucas, *Hunt and Working Terriers*, Chapman & Hall, 1931

Sir Peter J. Mackie, *The Keeper's Book: A Guide to the Duties of a Gamekeeper*, G. T. Foulis, revised edition 1929

Wickham Malins, *Bully & the Badger*, Robert Yeatman, 1974

Margaret Meek, 'The Limits of Delight', in Chris Powling, ed., *The Best of Books for Keeps: Highlights from the Leading Children's Book Magazine*, Bodley Head, 1994

Adrian Middleton and Richard Paget, *Badgers of Yorkshire and Humberside*, Ebor Press, 1974

Mr J. G. Millais, *Mammals of Great Britain*, vol. II, Longmans, 1904

Rosie Miller, *Are You a Badger or a Doormat?: How to Be a Leader Who Gets Results*, Prentice Hall (Financial Times), 2009

Ernest Neal, *The Badger*, Collins, 1948

Ernest Neal, *The Badger Man*, Providence Press, 1994

Ernest Neal and Chris Cheeseman, *The Badger*, T. & A. D. Poyser, 1996

Jan Needle, *Wild Wood*, Scholastic Press, 1990

Nimrod [Charles Apperley], *The Life of a Sportsman* [1832], John Lehmann, 1948

Barry F. Peachey, *Hunting the Badger*, Shooting News, 1993

George E. Pearce, *Badger Behaviour, Conservation and Rehabilitation*, Pelagic Publishing, 2011

Jeffrey Pearson, *Badger Bait*, Hessle Press, 1995

David Perkins, *Romanticism and Animal Rights*, Cambridge University Press, 2003

Frances Pitt, *Diana, My Badger*, Arrowsmith, 1929

Beatrix Potter, *The Tale of Mr. Tod*, Frederick Warne, 1912

Alison Prince, *Kenneth Grahame: An Innocent in the Wild Wood*, Allison & Busby, 1994

Jane Ratcliffe, *Through the Badger Gate*, G. Bell & Sons, 1974

Jane Ratcliffe, *Fly High, Run Free*, Chatto & Windus, 1979

Jane Ratcliffe, *Wildlife in My Garden*, Cicerone Press, 1986

Jacob Robinson and Sidney Gilpin, *North Country Sports and Pastimes. Wrestling and Wrestlers*, Bemrose & Sons, 1893

Theodore Roosevelt, *African Game Trails*, Charles Scribner's Sons, 1910

Timothy Roper, *Badger*, HarperCollins, 2010

Tim Sands, *Wildlife in Trust*, The Wildlife Trusts, 2012

Peter Savill et al., *Wytham Woods: Oxford's Ecological Laboratory*, Oxford University Press, 2011

Ernest Shepard, *Ben and Brock*, Methuen, 1965

Sylvia Shepherd, *Brocky*, Longmans, 1964

Eric Simms, *Voices of the Wild*, Putnam, 1957

Paul Skinner, Don Jefferies and Stephen Harris, *Badger Persecution and the Law*, The Mammal Society, 1989

Eileen Soper, *When Badgers Wake*, Routledge & Kegan Paul, 1955

Eileen Soper, *Eileen Soper's Badgers*, Robinson Publishing, 1992

Dr Geoffrey Sparrow, *The Terrier's Vocation*, Combridge, 1949

Adam Spencer, 'One Body of Evidence, Three Different Policies: Bovine Tuberculosis Policy in Britain', *British Journal of Politics and International Relations*, 31 (2), 2011

David Stephen, *Guide to Watching Wild Life*, Collins, 1963

David Stephen, *Bodach the Badger*, Century Publishing, 1983

J. C. Tregarthen, *The Life Story of a Badger*, John Murray, 1925

Ruthven Tremain, *The Animals' Who's Who*, Routledge and Kegan Paul, 1982

George Turbervile, *The Noble Art of Venerie or Hunting . . .*, 1576

Kathryn Wilkinson, 'Evidence Based Policy and the Politics of Expertise: A Case Study of Bovine TB', CRE Discussion Paper 12, 2007, http://www.ncl.ac.uk/cre/publish/discussionpapers/pdfs/dp12%20Wilkinson.pdf

Henry Williamson, 'The Badger Dig', in Henry Williamson, *The Village Book*, Jonathan Cape, 1930

Henry Williamson, *The Story of a Norfolk Farm*, Faber & Faber, 1941

Angus Wilson, *The Old Men at the Zoo*, Secker & Warburg, 1961

G. Wilson, S. Harris and G. McLaren, *Changes in the British Badger Population 1988 to 1997*, People's Trust for Endangered Species, 1997, http://jncc.defra.gov.uk/page-2797

INDEX